W9-CHJ-912

ISO 9000

QUALITY AND RELIABILITY

A Series Edited by

1. Designing for Minimal Maintenance Expense: The Practical Application of Reliability and Maintainability, *Marvin A. Moss*
2. Quality Control for Profit: Second Edition, Revised and Expanded, *Ronald H. Lester, Norbert L. Enrick, and Harry E. Mottley, Jr.*
3. QCPAC: Statistical Quality Control on the IBM PC, *Steven M. Zimmerman and Leo M. Conrad*
4. Quality by Experimental Design, *Thomas B. Barker*
5. Applications of Quality Control in the Service Industry, *A. C. Rosander*
6. Integrated Product Testing and Evaluating: A Systems Approach to Improve Reliability and Quality, Revised Edition, *Harold L. Gilmore and Herbert C. Schwartz*
7. Quality Management Handbook, *edited by Loren Walsh, Ralph Wurster, and Raymond J. Kimber*
8. Statistical Process Control: A Guide for Implementation, *Roger W. Berger and Thomas Hart*
9. Quality Circles: Selected Readings, *edited by Roger W. Berger and David L. Shores*

10. Quality and Productivity for Bankers and Financial Managers, *William J. Latzko*
11. Poor-Quality Cost, *H. James Harrington*
12. Human Resources Management, *edited by Jill P. Kern, John J. Riley, and Louis N. Jones*
13. The Good and the Bad News About Quality, *Edward M. Schrock and Henry L. Lefevre*
14. Engineering Design for Producibility and Reliability, *John W. Priest*
15. Statistical Process Control in Automated Manufacturing, *J. Bert Keats and Norma Faris Hubele*
16. Automated Inspection and Quality Assurance, *Stanley L. Robinson and Richard K. Miller*
17. Defect Prevention: Use of Simple Statistical Tools, *Victor E. Kane*
18. Defect Prevention: Use of Simple Statistical Tools, Solutions Manual, *Victor E. Kane*
19. Purchasing and Quality, *Max McRobb*
20. Specification Writing and Management, *Max McRobb*
21. Quality Function Deployment: A Practitioner's Approach, *James L. Bossert*
22. The Quality Promise, *Lester Jay Wollschlaeger*
23. Statistical Process Control in Manufacturing, *edited by J. Bert Keats and Douglas C. Montgomery*
24. Total Manufacturing Assurance, *Douglas C. Brauer and John Cesarone*
25. Deming's 14 Points Applied to Services, *A. C. Rosander*
26. Evaluation and Control of Measurements, *John Mandel*
27. Achieving Excellence in Business: A Practical Guide to the Total Quality Transformation Process, *Kenneth E. Ebel*
28. Statistical Methods for the Process Industries, *William H. McNeese and Robert A. Klein*
29. Quality Engineering Handbook, *edited by Thomas Pyzdek and Roger W. Berger*
30. Managing for World-Class Quality: A Primer for Executives and Managers, *Edwin S. Shecter*
31. A Leader's Journey to Quality, *Dana M. Cound*
32. ISO 9000: Preparing for Registration, *James L. Lamprecht*
33. Statistical Problem Solving, *Wendell E. Carr*
34. Quality Control for Profit: Gaining the Competitive Edge. Third Edition, Revised and Expanded, *Ronald H. Lester, Norbert L. Enrick, and Harry E. Mottley, Jr.*
35. Probability and Its Applications for Engineers, *David H. Evans*
36. An Introduction to Quality Control for the Apparel Industry, *Pradip V. Mehta*
37. Total Engineering Quality Management, *Ronald J. Cottman*
38. Ensuring Software Reliability, *Ann Marie Neufelder*
39. Guidelines for Laboratory Quality Auditing, *Donald C. Singer and Ronald P. Upton*

ISO 9000

PREPARING FOR REGISTRATION

JAMES L. LAMPRECHT

International Consultant
Scottsdale, Arizona

ASQC Quality Press — Milwaukee

Marcel Dekker, Inc. — New York • Basel • Hong Kong

Library of Congress Cataloging-in-Publication Data

Lamprecht, James L.
ISO 9000 : preparing for registration / James L. Lamprecht.
p. cm.— (Quality and reliability ; 32)
Includes bibliographical references and index.
ISBN 0-8247-8741-2 (acid-free paper)
1. Quality assurance—Management. 2. Quality control—Standards.
I. Title. II. Series.
TS156.6.L36 1992
658.5'62—dc20 92-4932
CIP

This book is printed on acid-free paper.

ASQC
Quality Press
611 East Wisconsin Avenue
Milwaukee, Wisconsin 53202

Marcel Dekker, Inc.
270 Madison Avenue
New York, New York 10016

Current printing (last digit):

10 9 8 7

PRINTED IN THE UNITED STATES OF AMERICA

To my parents

About the Series

The genesis of modern methods of quality and reliability will be found in a simple memo dated May 16, 1924, in which Walter A. Shewhart proposed the control chart for the analysis of inspection data. This led to a broadening of the concept of inspection from emphasis on detection and correction of defective material to control of quality through analysis and prevention of quality problems. Subsequent concern for product performance in the hands of the user stimulated development of the systems and techniques of reliability. Emphasis on the consumer as the ultimate judge of quality serves as the catalyst to bring about the integration of the methodology of quality with that of reliability. Thus, the innovations that came out of the control chart spawned a philosophy of control of quality and reliability that has come to include not only the methodology of the statistical sciences and engineering, but also the use of appropriate management methods together with various motivational procedures in a concerted effort dedicated to quality improvement.

This series is intended to provide a vehicle to foster interaction of the elements of the modern approach to quality, including statistical applications, quality and reliability engineering, management, and motivational aspects. It is a forum in which the subject matter of these various areas can be brought together to allow for effective integration of appropriate techniques. This will promote the true benefit of each, which can be achieved only through their interaction. In this sense, the whole of quality and reliability is greater than the sum of its parts, as each element augments the others.

The contributors to this series have been encouraged to discuss fundamental concepts as well as methodology, technology, and procedures at the leading edge of the discipline. Thus, new concepts are placed in proper perspective in these evolving disciplines. The series is intended for those in manufacturing, engineering, and marketing and management, as well as the consuming public, all of whom have an interest and stake in the improvement and maintenance of quality and reliability in the products and services that are the lifeblood of the economic system.

The modern approach to quality and reliability concerns excellence: excellence when the product is designed, excellence when the product is made, excellence as the product is used, and excellence throughout its lifetime. But excellence does not result without effort, and products and services of superior quality and reliability require an appropriate combination of statistical, engineering, management, and motivational effort. This effort can be directed for maximum benefit only in light of timely knowledge of approaches and methods that have been developed and are available in these areas of expertise. Within the volumes of this series, the reader will find the means to create, control, correct, and improve quality and reliability in ways that are cost effective, that enhance productivity, and create a motivational atmosphere that is harmonious and constructive. It is dedicated to that end and to the readers whose study of quality and reliability will lead to greater understanding of their products, their processes, their workplaces, and themselves.

Edward G. Schilling

Preface

Over the past twelve to eighteen months, an increasing number of businesses across America have begun receiving memos or questionnaires requesting information relating to the supplier's anticipated date of International Standard Organization (ISO) 9000 registration. This material has come from various sources — domestic and foreign customers, corporate headquarters, and overseas subsidiaries or sales representatives. In some cases, suppliers who have already experienced ISO 9000-type audits from their customers have been told to "shape up" within the next twelve to eighteen months.

Often uncertain as to what the International Standards Organization 9000 series is all about, countless CEOs, quality directors, engineers and others have been crowding ISO 9000 seminars in the hope of finding out more about this mysterious acronym. Returning from the seminar, attendees often realize that they still don't know how to address the many facets of the International Standards Organization 9000 series. This book is intended for them. More specifically, it is addressed to those who wish to organize, document, or otherwise implement a quality assurance system based on the ISO 9000 (American National Standards Institute/American Society of Quality Control Q90-Q94) series of quality standards.

The book evolved from a series of seminars first offered in Europe in the fall of 1989 and in the U.S. starting in early 1991. Motivated by the success of the seminars, the author began reorganizing the seminar in a book format in the spring of 1991. The book's primary objectives are to:

- Inform the reader about the five ISO standards.

- Review and interpret each of the twenty paragraphs of the ISO 9001/Q91 standard (the most comprehensive quality system).

- Present various scenarios on what to consider and how to approach registration (i.e., what to register?).

- Present a documentation model designed to address the ISO 9001 (as well as 9002 or 9003) quality assurance system.

- Offer suggestions on how to bring about change, plan, organize or otherwise delegate the ISO implementation process.

- Give some insights as to the various registrars currently operating in the U.S.

- Review the *Guidelines for Auditing Quality Systems* as presented in the ISO 100011-1, 2, and 3 series.

- Provide a preview of things to come (1992, 1996 editions of ISO).

In addition, suggestions are also offered regarding timing, planning, cost, self-assessment (an ISO 9000 questionnaire is included in Appendix C), and other important questions relating to registration, third party and internal audits, and registrars in general.

Although a myriad of training seminars and conferences are currently offered by various organizations on what the ISO 9000 series is, is not, or should be, to the author's knowledge, no book currently exists on how to facilitate and implement the ISO 9000 series of standards. It is hoped that the information, examples, and suggestions presented here will facilitate the implementation task awaiting the many quality managers or ISO committees in charge of the ISO 9000 project.

My purpose in writing this book was not to compare the ISO 9000 series of standards with other standards such as, MIL-Q-9858-A or

MIL-I-45208A (there are in fact many similarities between the two sets of standards). Nor was my intent to contrast the ISO series with the many Supplier Quality Assurance (SQA) systems developed over the years. Organizations familiar with their customers' SQAs and other industrial or governmental standards are no doubt better informed about the requirements imposed by these standards than the author. Finally, although occasional references are made to the Malcolm Baldrige Award and other national awards, my intent is not to compare and contrast the ISO 9000 series of standards with national awards. My reluctance to do so is motivated by the fact that all too often such debates quickly turn into unproductive emotional or, at times, nationalistic arguments and counterarguments (see, for example, the May and August 1991 issues of *Quality Progress*). I would hope that after reading the book, individuals having expertise with a particular set of standards will be able to better compare and contrast the ISO 9000 series with those standards.

The book is not intended to be a substitute for the ANSI/ASQC Q90-Q94 (ISO 9000-9004) series. In fact, purchase of these standards from either the American National Standards Institute or the American Society of Quality Control is highly recommended as a companion to this book. Purchase of the ISO 10011-1: *Auditing* 10011-2: *Qualification criteria for quality systems auditors*, and 10011-3: *Management of Audit Programs* series is also recommended.

While addressing the subject of ISO registration I have, on several occasions, gone beyond the simple mechanics of registration and addressed some of the more general issues of quality management. In doing so however, I have deliberately focused my attention and limited my comments as they relate to ISO 9004 (ANSI/ASQC Q94-1987) *Quality Management and Quality System Elements — Guidelines* and ISO 9000 (ANSI/ASQC Q90-1987) *Quality Management and Quality Assurance Standards — Guidelines for Selection and Use.* Even though one of the primary objectives of this book is to present an approach on how to document a quality assurance system which will comply to any one of the three ISO models of quality assurance (9001, 9002, or

9003), I believe that reference to the two key supporting documents (9000/Q90 and 9004/Q94) is essential if one is to understand ISO's *modus operandi.* Having said this, I should emphasize that this book is not one more revelation on how to implement a so-called Total Quality Management system (see Chapter 1 for additional comments). Readers wishing to obtain more information on the subject of Total Quality Management are referred to the plethora of books already available on the subject.

Chapter 1 is an overview on what the ISO 9000 series is all about, its scope and field of application. Chapter 2 reviews each of the twenty clauses of ISO 9001. Chapter 3 offers a look at how to interpret the standard's jargon. Chapters 4 through 7 address the issues of organization and how to write documentation. Examples are provided throughout these chapters. Chapter 8 offers suggestions on how to prepare for the implementation efforts. Time and cost estimations are also provided. Chapter 9 answers some of the most often asked questions about the cost of implementing the ISO series. Chapter 10 addresses some often asked questions about the standards. Chapter 11 looks at the qualification requirements of the registrars. Chapter 12 reviews the third-party audit process. Chapter 13 covers some of the most important issues regarding internal audits. Chapter 14 looks at possible updates to the ISO series in the year 1992 and beyond. Chapter 15 concludes with some general comments and observations regarding ISO series, ISO and the service industry, and consultation in general.

I hope you will find the information contained herewith of value. Let me know what you think. Your suggestions and *constructive* criticism on how to improve are earnestly sought.

Acknowledgments

This book could not have evolved without the many contributions offered by the seminar's participants. Their challenging questions, debates, arguments, and interpretations constantly forced the author to go back to the Standards and Guidelines to read and re-read the many paragraphs, as well as call fellow auditors or survey registrars for their interpretation or experience with similar questions. I would also like to express my thanks to the anonymous reviewers; their comments were most helpful.

Thanks are due to the efficient and responsive staff of Marcel Dekker, Inc., whose professionalism and dedication played an important role in the timely completion of this project. I would like to thank my wife, Shirley, for supporting me throughout the duration of this seemingly endless project.

James L. Lamprecht

Table of Contents

1 Introduction

January 1, 1993 will no doubt be an important day for the European Community member countries for it will mark a new beginning in Pan-European trade. Although difficult crossroads still lay ahead, harmonization is well on its way and has already affected the way American companies conduct business in Europe. As they enter the 1990s, American multinationals face a host of challenges. Of particular relevance to the future of U.S.-European trade and the well-being of America's competitiveness are: (1) the ***International Standards Organization (ISO) 9000-9004*** series of standards (known in the U.S. as the ANSI/ASQC Q90-Q94 series), a series of documents on quality assurance rapidly gaining recognition in this country; and (2) the ability of companies to successfully respond to the challenge by rapidly implementing an ISO 9000 type quality system.

What Is the ISO 9000 Series?

Occasionally perceived in this country as a European standard, the ISO 9000/Q90 series of standards is in fact an international set of documents written by members of a worldwide delegation known as the ISO/Technical Committee 176. The ISO/TC 176 consists of three standards committees and several working groups. Four national associations participate in the ISO/TC 176 as convenors of subcommittees, they are: AFNOR (Association Française de Normalisation), ANSI (American National Standards Institute), BSI (British Standards Institute), NNI (Nederlands Normalisatie Instituut), and SCC (Standards Council of Canada). Other member countries are also represented by their respective national standards bodies.

Primarily conceived to help harmonize the large number of national and international standards (see Table 1.1 for a listing of ISO 9000 equivalent standards adopted by National Standards Bodies), the standards are intended to be used in contractual and non-contractual situations. As the *Guidelines for Selection and Use* explains, "In both

Table 1.1 National Standards Equivalent to the ISO 9000 Series

Standards Body	Quality Management and quality assurance standards: Guidelines for selection and use	Quality Systems model for QA in design/dvp. production, installation and servicing	Quality Systems: model for QA in production and installation	Quality Systems: model for QA in final inspection and test	Quality Mngt. and quality systems guidelines
ISO	ISO 9000:87	ISO 9001:87	ISO 9002:87	ISO 9003:87	ISO 9004:87
Australia	AS 3900	AS 3901	AS 3902	AS 3903	AS 3904
Austria	OE NORM-PREN 29000	OE NORM-PREN 29001	OE NORM-PREN 29002	OE NORM-PREN 29003	OE NORM-PREN 29004
Belgium	NBN X 50-0020-1	NBN X 50-003	NBN X 50-004	NBN X 50-005	NBN X 50-002-2
Canada	-	-	-	-	CSA Q420-87
China	GB/T 10300.1-88	GB/T 10300.2-88	GB/T 10300.3-88	GB/T 10300.4-88	GB/T 10300.5-88
Denmark	DS/EN 29000	DS/EN 29001	DS/EN 29002	DS/EN 29003	DS/EN 29004
European	EN 29000:87	EN 29001:87	EN 29002:87	EN 29003:87	EN 29004:87
Finland	SFS-ISO9000	SFS-ISO9001	SFS-ISO9002	SFS-ISO9003	SFS-ISO9004
France	NF X 50-121	NF X 50-131	NF X 50-132	NF X 50-133	NF X 50-122
Hungary	MI 18990-88	MI 18991-88	MI 18992-88	MI 18993-88	MI 18994-88
India	IS:10201 Part 2	IS: 10201 Part 4	IS: 10201 Part 5	IS: 10201 Part 6	IS: 10201 Part3
Ireland	IS 300 Part 0	IS 300 Part 1	IS 300 Part 2	IS 300 Part 3	IS 300 Part 0
Italy	UNI/EN 29000-1987	UNI/EN 29001-1987	UNI/EN 29002-1987	UNI/EN 29003-1987	UNI/EN 29004-1987
Malaysia	-	MS 985/ ISO 9001:87	MS 985/ ISO 9002:87	MS 985/ ISO 9003:87	-
Netherlands	NEN-ISO 9000	NEN-ISO 9001	NEN-ISO 9002	NEN-ISO 9003	NEN-ISO 9004
New Zealand	NZS 5600: Part 1	NZS 5601-87	NZS 5602-87	NZS 5603-87	NZS 5600: Part 2-1987
Norway	NS-EN 29000	NS:EN 29001	NS-ISO 9002	NS-ISO 9003	-
South Africa	SABS 0157: Part 0	SABS 0157: Part I	SABS 0157: Part II	SABS 0157: Part III	SABS 0157: Part IV
Spain	UNE 66 900	UNE 66 901	UNE 66 902	UNE 66 903	UNE 66 904
Sweden	SS-ISO 9000	SS-ISO 9001	SS-ISO 9002	SS-ISO 9003	SS-ISO 9004
Switzerland	SN-ISO 9000	SN-ISO 9001	SN-ISO 9002	SN-ISO 9003	SN-ISO 9004
Tunisia	NT 110.18	NT 110.19	NT 110.20	NT 110.21	NT 110.22
U. Kingdom	BS 5750: Part 0 Section 0.1	BS 5750: Part 1	BS 5750: Part 2	BS 5750: Part 3	BS 5750: Part 4
USA	ANSI/ASQC Q90	ANSI/ASQC Q91	ANSI/ASQC Q92	ANSI/ASQC Q93	ANSI/ASQC Q94
USSR	-	40.9001-87	40.9002.87	-	-
Germany	DIN ISO 9000	DIN ISO 9001	DIN ISO 9002	DIN ISO 9003	DIN ISO 9004
Yugoslavia	JUS A.K. 1.010	JUS A.K. 1.012	JUS A.K. 1.013	JUS A.K. 1.014	JUS A.K. 1.011

situations, the supplier's organization wants to install and maintain a quality system that will strengthen its own competitiveness and achieve the needed product quality in a cost effective way" (ANSI/ASQC Standard Q90-1987, p. 2).

Architecture of the ISO 9000 Series of Standards

The ISO 9000 series consists of five documents: three core quality system documents which are models of quality assurance, namely 9001/Q91, 9002/Q92 and 9003/Q93 and two supporting guidelines documents, 9000/Q90 and 9004/Q94. The documents' title clearly indicate their purpose.

ISO 9000 (ANSI/ASQC Q90-1987)
Quality Management and Quality Assurance Standards: Guidelines for Selection and Use.

ISO 9001 (ANSI/ASQC Q91-1987)
Quality Systems—Model for Quality Assurance in Design/Development, Production, Installation, and Servicing.

ISO 9002 (ANSI/ASQC Q92-1987)
Quality Systems—Model for Quality Assurance in Production and Installation.

ISO 9003 (ANSI/ASQC Q93-1987)
Quality Systems—Model for Quality Assurance in Final Inspection and Test.

ISO 9004 (ANSI/ASQC Q94-1987)
Quality Management and Quality System Elements Guidelines.

Whereas the three models for quality assurance "represent three distinct forms of functional or organizational capability suitable for two-party contractual purposes" (ANSI/ASQC Q91-1987, p. 1), the ISO

9000/Q90 and the ISO 9004/Q94 standards are nothing more than guidelines. The ISO 9000/Q90 set of guidelines were primarily written to help potential users (both customers/purchasers and suppliers) decide which quality assurance model is most appropriate and relevant to a particular contractual relationship. The ISO 9004/Q94 provides its user with a set of guidelines "by which quality management systems can be developed and implemented" (ANSI/ASQC Q94, p. 2). It is important to note that *neither* the Q90 *nor* the Q94 standard is a model of quality assurance. Consequently, Q90 and Q94 should not be perceived as a set of mandatory requirements. [Note: Since ISO 9000/Q90 and ISO 9004/Q94 are guidelines, it is in fact erroneous to speak of "ISO 9000 or Q90 registration/certification." One can only obtain registration to either 9001, 9002 or 9003.]

The series of standards (9000-9004), are *not technical in content* in that they do not specify nor set criteria — i.e., minimum purity, pH tolerances, hardness requirements, etc., for products. Rather, the quality system standard 9001/Q91, 9002/Q92 and 9003/Q93, "complement relevant product or service requirements given in the technical specifications" (ANSI/ASQC Standard Q90-1987, p. 1). Each standard focuses on the documentation of *operational techniques and managerial activities used to fulfill customer expectations and requirements.*

Field and Scope of Application

Contrary to what is occasionally believed, the three ISO models for quality assurance (9001, 9002 and 9003), were not written for any specific industry. Although some industries have issued their own set of interpretative guidelines (see for example the U.K.'s Chemical Industries Association ISO 9001-EN 29001 *Guidelines for Use by the Chemical and Allied Industries*), these guidelines provide little, if any, additional information. The fact remains that the standards are generic and are intended to apply to *all* industries.

Depending on your point of view, the generic nature of the ISO standards can be perceived as a source of wisdom or confusion. Since the standards do not dictate how you should implement the requirements but rather focuses on the issues of how you address each requirement, one could argue that such a magnanimous approach is indeed wise. Unfortunately, the generic nature of the standards simultaneously leads to interpretive difficulties. Fortunately, the ISO/TC 176 committee, well aware of the standards' intent to cover a broad spectrum of industries, does recognize that tailoring might be required to best suit each individual need.

> *It is intended that these Standards will normally be adopted in their present form,* ***but on occasions they may need to be tailored for specific contractual situations*** [ANSI/ASQC Standards Q91-1987, p.1].

Recognizing that at the heart of most customer-supplier relationships are contractual obligations, the international committee chose to organize the ISO quality assurance documents into three broad categories: 9001, 9002 and 9003.

ISO 9001 is applicable in contractual situations when "conformance to specified requirements is to be assured during several stages which *may* include design/development, production, installation, and servicing" [ANSI/ASQC Standard Q91, p. 1, emphasis added]. Moreover,

a) The contract requires design effort and the product requirements are stated principally in performance terms or they need to be established;

b) Confidence in product conformance can be attained by adequate demonstration of certain supplier's capabilities in design, development, production, installation, and servicing (ANSI/ASQC Standard Q91-1987, p. 1).

ISO 9002, the standard often adopted (in the U.S. and abroad) by the chemical and process industries, as well as other industries, is applicable in contractual situations when:

a) The specific requirements for the product are stated in terms of an established design or specification;

b) Confidence in product conformance can be attained by adequate demonstration of a certain supplier's capabilities in production and installation (ANSI/ASQC Standard Q92-1987, p. 1).

ISO 9003, which has been applied to small shops, distributors or divisions within an organization, is applicable in contractual situations when:

The conformance of the product to specified requirements can be shown with adequate confidence providing that certain supplier's capabilities for inspection and tests conducted on the product supplied can be satisfactorily demonstrated on completion (ANSI/ASQC Standard Q93-1987, p.1).

This partitioning along contractual obligations is both logical and practical. For instance, whereas ISO 9003 suppliers only have to demonstrate, among other things, a *capability* to inspect and test a product (e.g., using proper acceptance sampling plans, monitoring test equipments, etc.), ISO 9002 suppliers are expected to go further and demonstrate that the relevant manufacturing processes are capable of maintaining requirements as per design specifications (e.g., perform machine and/or process capability studies). ISO 9001 suppliers have to go further yet, and must *also* demonstrate a capability/mastering in *all* phases of design, development and servicing.

Table 1.2 Cross-Reference List of Quality System Elements
(Adapted from the Annex found in ISO 9000, p. 6)

Title	Corresponding Paragraph (or subsection) Nos. in		
	9001	**9002**	**9003**
Management Responsibility	4.1	4.1a	4.1b
Quality System Principles	4.2	4.2	4.2a
Contract Review	4.3	4.3	-
Design Control	4.4	-	-
Document Control	4.5	4.4	4.3a
Purchasing	4.6	4.5	-
Purchaser Supplier Product	4.7	4.6	-
Product Identification and Traceability	4.8	4.7	4.4a
Control of Production	4.9	4.8	-
Inspection and Testing	4.10	4.9	4.5a
Inspection, Measuring and Test Equip.	4.11	4.10	4.6a
Inspection and Test Status	4.12	4.11	4.7a
Control of Nonconforming Product	4.13	4.12	4.8a
Corrective Action	4.14	4.13	-
Handling, Storage, Packaging and Delivery	4.15	4.14	4.9a
Quality Records	4.16	4.15	4.10a
Internal Audits	4.17	4.16a	-
Training	4.18	4.17a	4.1b
After-sales Servicing	4.19	-	-
Statistical Techniques	4.20	4.18	4.12a

a **Less stringent than 9001**
b **Less stringent than 9002**
- **Element not present**
Unmarked paragraphs indicate Full Requirement

What Model to Choose (9001, 9002 or 9003)?

When an organization commits to achieving registration to one of the three ISO standards it must first decide (unless otherwise dictated by its customer(s)), which one of the three models of quality assurance (9001, 9002 or 9003) best suits its needs. As indicated in Table 1.2, 9003 is a subset of 9002 which in turn is a subset of 9001. The most detailed model for quality assurance, 9001/Q91, consists of twenty paragraphs. The 9002/Q92 standards consists of eighteen paragraphs whereas 9003/Q93 only has twelve. When one recalls that 9003/Q93 is the least used model, the decision as to which standards best suits a company's need should be rather straightforward. In some cases, the decision tree consists of a simple binary choice: "Are we involved with Design and Development?" If the answer is "yes," then you should choose 9001/Q91. If the answer is "no," then 9002/Q92 is very likely to suit your needs. Unfortunately, for reasons that will be explored in Chapter 3, the decision as to which model to follow is not always straightforward.

Who Should Implement an ISO 9000 Quality System?

The range of businesses that can benefit from an ISO (9001, 9002 or 9003) type quality assurance system is very broad. Multinationals are rapidly discovering that they often have little choice but to adopt the standard in order to become global competitors. In an increasing number of cases, registration to one of the ISO standards is becoming mandatory. The diffusion of ISO awareness across the spectrum of American industry seems to follow a particular pattern. The early stages (pre-1990) saw multinationals register some of their key plants (i.e., plants whose market relied heavily on exports). This process continues today but with a new added feature. Indeed, as more and more companies become registered to one of the ISO quality assurance systems (over eighty-five percent of all companies choose either 9001 or 9002 as one of their models for quality assurance), they immediately request their suppliers to address the ISO 9000 registration issue. Consequently, as of mid-1991, more and more small to medium size suppliers are being asked by their customers to achieve, in "the near future," registration to either 9001, 9002 or 9003. Why more and more suppliers are asked to obtain "ISO registration" (or certification, as it is often referred to), will be explored below.

There are of course other factors. The Department of Defense — prompted by NATO — decided in August 1989 to adopt the ISO Standards. Although the DOD will eventually replace its MIL-Q-9858A and MIL-I-45028A with the ISO 9001 standard, it does not (yet) require its supplier to be registered to the ISO 9001. Nonetheless, as far as DOD suppliers are concerned, the motivation to achieve ISO certification will no doubt remain high.

Of particular interest to the U.S. aerospace industry is the fact that the Society of British Aerospace Companies, Inc. (SBAC), has launched an ISO 9000 registration scheme for the United Kingdom Aviation Industry. Similarly, members of the commercial aviation industry should be interested in learning that the European aviation regulatory system (JAA) does recognize the ISO standards. Ever since airworthiness codes have been finalized, Europe's Joint Aviation Regulation (JAR) 145 (Maintenance Organization Approval) immediately made reference to the ISO standards. Other industries including the pharmaceutical, electronic-telecommunications, construction, petroleum and related industries, as well as countless others will sooner or later be affected by the ISO standards. Consequently, in view of these events, U.S. companies will *likely* have to achieve ISO registration if they either:

(1) Export to one or more of the twelve member countries of the European Economic Community (EEC) or the European Free Trade Association (EFTA). [Note: EFTA members include: Iceland, Norway, Sweden, Switzerland, Liechtenstein, Finland and Austria. Sweden and Austria have applied for EEC membership. Eastern European countries plan to join within the next five to eight years.]

(2) If any of their customers export to the EEC, EFTA or supply the DOD — this is often referred to as the "cascading effect" whereby customers request certification of their suppliers who in turn request it of their suppliers and so on down the line. Note that although the DOD does not require ISO 9001 registration, this does not ensure that suppliers to the DOD will not perceive ISO 9001 to be a necessary prerequisite to do business. Such (un)fortunate misconceptions or rumors about what *is* or *is not* required have already reached out of control proportions.

(3) If they are a subsidiary of a European company. In such cases, companies have already achieved ISO registration or are well on their way toward registration.

Finally, one should observe that the "requirement" for ISO 9000 registration is not geographically limited to Europe. Some U.S. firms have had to acquire ISO 9000 registration to satisfy their Middle Eastern and Asian customers. Similarly, plants in Singapore, Argentina, Brazil and Mexico are currently achieving ISO 9001 or 9002 registration.

ISO 9000: Trade Barrier or Opportunity to Improve?

The ISO 9000 series of standards is not likely to be used in the near future as an entry barrier by European Community (EC) member countries. My experience with one California based telecommunications company which had successfully bid on an important British contract, despite the fact that it was not fully compliant with ISO 9000, would *partly* disprove such claims. I emphasize *partly* because in this particular case, the driving force which led to the successful bid was not solely linked to the ISO 9000 registration process but also involved a series of important factors not least of which was: (1) the participation during all phases of the proposal of a Canadian firm *which had received ISO 9000 registration* and (2) the competitively priced technical know-how of the American firm. One should not infer however that ISO 9000 compliance was not specified in the original contract, *for indeed it was* in paragraph 3.5 of the "Basic Quality Assurance Requirements," which specified:

> *The contractor is free to choose the system and processes by which he is best able to meet, and demonstrate that he meets, the client's (name of company withheld) requirements. However the client's Quality Assurance Division Representative will use the appropriate part of British Standards 5750: 1987 (ISO 9000 et seq.) as a basis for judging the adequacy and effectiveness of the system used in supplying the goods and services through the duration of the contract.*

Since the California vendor was responsible for the design/development, production, installation and servicing of the telecommunications equipment, ISO 9001 applied.

Whereas it is true that as of late 1991 ISO certification had not been used as an entry barrier to the European market by European governments, it should be emphasized that some U.S. multinationals have already been told by a few of their major European customers that they could not be included on their list of preferred suppliers unless they were registered to one of the ISO standards!

When faced with such contractual requirements, the reaction is often predictable and usually consists of several fundamental questions: "What is ISO 9000?" , "Would I pass such an audit?" and if not, "How quickly can I get ISO ready?" Deciding whether one is or is not in compliance with any of the many paragraphs of the ISO 9001, 2 or 3 series can be a difficult and, at times, confusing task. Although not in the spirit of a *Total Quality Commitment* but for the sake of practical expediency, some firms in Europe and in the U.S. simply hire a consulting firm to help guide them towards ISO 9000 certification. Others see the ISO 9000 series as a wonderful opportunity to adopt their favored Total Quality Management (TQM) policy. In the case of the California firm, senior management chose to avoid any mishaps and opted for an ISO 9001 pre-audit.

Clarifying Some Misunderstandings About the ISO Standards

At the risk of being redundant, it is important to underline that the ISO 9000 series of standards is NOT a European standard nor should it be referred to as "the ISO specifications." The American National Standards Institute (ANSI) is an active participating member on ISO working groups. Consequently, American opinions are well represented in ISO. Naturally, this does not (nor can it ever) ensure that all major sectors of the American industrial complex are represented.

Another common mistake, often made by the uninitiated, is to refer to the registration process as "getting ISO 9000 certification." This unfortunate habit, which has now become part of the unofficial jargon, has caused a great deal of confusion. As was explained in pages four and five, the ISO 9000 document is nothing more than a *Guidelines for Selection and Use.* Since it is not one of the models for quality assurance, one cannot achieve registration to ISO 9000. Moreover, rather than speaking of "certification" or "ISO certification," it is preferable to refer to the *registration process* (to be discussed in Chapters 11-12) [In the U.K., the registration process is referred to as *accreditation.*] Certification implies to many people *product certification*, which has *nothing* to do with the ISO 9000 series. This leads to yet another misconception about the ISO series.

Since the ISO 9000 series of standards does not set any specifications (product or otherwise), one should not refer to the ISO 9000 series as a set of specifications. This confusion might partly be due to the fact that some people perceive the ISO 9000 series as a set of documents which affects or otherwise complements product specification. Such is not the case. It is an international standard for the implementation and management of a quality assurance system. As such it is not a technical document, but rather a *generic* document intended to apply to *all* industries. This is clearly stated in the opening paragraphs of all five ISO documents. For example, one of the opening paragraphs of 9001/Q91 states:

> *It is* ***emphasized*** *that the quality system requirements specified in this Standard, Standards Q92 (ISO 9002) and Q93 (ISO 9003) are* ***complementary*** *(not alternative) to the technical (product/service) specified requirements (emphasis added) [ANSI/ASQC Standard Q91-1987, p. 1].*

Another common myth about the ISO 9000 series is that it will be enforced by the European Community on January 1st, 1993. Nothing can be further from the truth. Central to the drive to develop a uniform set of quality assurance systems was the European Community Product Directive Liability (July 1985). This directive was amended in 1989 to include a Product Safety Directive which requires the implementation of a quality system (hence the need for an ISO 9000 type quality assurance system). As W.H. Boehling explains in the June 1990 issue of *Quality Progress*, these directives would obligate manufacturers and importers to permanently monitor the safety of their products (see also Chapter 15 for a list of product categories).

Although it is true that the European Community will theoretically achieve instantaneous economic unity, one should not necessarily assume that doomsday will occur for all companies which have not achieved ISO certification by midnight on December 31, 1992. Indeed, as the following excerpts from *The Economist* of June 22, 1991 explains, much work will still have to be done.

> *At first glance the European Community's program of 282 measures to remove internal barriers by the end of 1992 appears well on the way to being realized: the Council of Ministers and the European Parliament have approved 198 laws so far. They take effect, however, only when member states pass corresponding legislation.*

Many governments treat this task casually: only Denmark, France and Holland, for instance, have bothered to implement a law on the mutual recognition of architects' diplomas. Of the 126 single-market laws that should by now have been implemented, only 37 have been passed by all 12 members. The best performers are Denmark, which has implemented 107, France with 103, Britain with 99 and Portugal with 96. Among the laggards, Italy has implemented only 52 laws, Ireland 74, Luxembourg 81 and Spain 83.

[The article concludes with some revealing examples of disharmony.]

Germany, with its perfectionist tradition of industrial standards, is one of the most frequent subjects of complaint. But when the commission last met German officials, 22 out of 30 outstanding cases were sorted out—including the Bundespost's refusal to approve British-built fax machines and satellite receivers that failed to meet German standards. In May the commission also asked Spain to stop blocking imports of computer keyboards. A regulation prevented those which lacked the letter "ñ" from being sold in Spain [Economist, June 22, 1991, p. 76].

Despite, or perhaps in spite of these interesting anecdotes there are, as has already been stated, many compelling reasons to start implementing an ISO quality assurance system. Indeed, if members of the EC keep its own commission busy arbitrating complaints, think of what might happen to non-EC countries who are not registered to one of ISO standards! Based on the variety of industry representatives who are currently (1991) attending the many seminars offered on ISO 9000 certification, one can safely conclude that the ISO series of quality systems is here to stay.

ISO and OSHA

When faced with the task of implementing an ISO quality assurance system, the quality manager or the local ISO 9000 committee will invariably be confronted with some pessimism on the part of some of their co-workers. The rationale often used by co-workers to justify the impracticality of ISO 9001 or 9002 is lack of time. Unfortunately, although time is everyone's nemesis, other factors might well require a company to implement an ISO quality system. Health and safety federal regulations for example, require many companies to implement a quality system that parallels and in some cases duplicates many of ISO's requirements (see for example the July 17, 1990 *Federal Register* Part III: Department of Labor Occupational Safety and Health Administration 29 CFR Part 1910 Process Safety Management of Highly Hazardous Chemicals; Notice of Proposed Ruling.)

ISO and Other Standards (American Petroleum Institute Q1 and the Federal Aviation Regulations)

On 18 December 1990, the ISO TC (Technical Committee) 67 met in Oslo to compare API Q1 and ISO 9001. The working group, which included delegates from France, Germany, Italy, Japan, the Netherlands, Norway and the U.K., came up with the following three resolutions:

> *Resolution 1*
>
> ISO TC/67 is advised to adopt the ISO 9000 series of standards for its applicability in the Oil and Gas business economics sector, thus following the principles of the ISO TC/176 report N141 "A strategy for international standards implementation in the quality arena during the nineties."

Resolution 2

ISO TC/67 is advised to discuss certification principles, on a global level, within the Oil and Gas business economic sector as a separate subject.

Resolution 3

The working group recommends that the API organization should be informed of the resolutions attached herewith.

The working group came to the conclusion that whereas API Q1 deals with license requirements and procedures as they relate to quality program requirements, ISO 9001 deals with quality system requirements in a contractual situation. A detailed comparison between ISO 9001 and API Q1 was undertaken by the ISO TC/67 working group.

API's reply to the ISO TC/67 Oslo report, dated February 20, 1991 concluded that:

> It appears that the Working Group (i.e., TC/67) is not familiar with the interrelationship of specific sections of API Spec Q1 within itself (e.g., Section 3.17.2 with Section 3.6.7) and the interrelationship of API Spec Q1 with the API product specification(s).
>
> Therefore, it is the opinion of the API committee 18 that the statements and conclusions of ISO/TC 67 Working Group 1 are incomplete and inaccurate and should not form the basis of any resolution.

The API's reluctance to accept ISO 9001 as a substitute for its own Spec Q1 is understandable. Much time, effort and energy has no doubt been invested by various members of the API and, as was clearly evidenced during the August 1, 1991 meeting in Houston, Texas, some members of the API committee were not about to have their Spec Q1 swept under the rug by ISO 9001. Instead, the API, to the chagrin of some of its members (particularly some of the non-voting manufacturers), tentatively agreed to "align" Spec Q1 with ISO 9001 to, in the words of one member, "make it better than ISO 9001." The desire to enhance or otherwise align the ISO 9000 series with other U.S. standards or regulatory agencies has been expressed by others. In a 1991 brochure advertising an upcoming Conference on Quality in Commercial Aviation, one could read the following abstract:

> In the United States, conversion (to ISO 9000) is complicated by the difference between the Federal Aviation Regulations (FAR), Part 21 or more specifically, sub parts F, G and K, AC-21-1B, AC-21-6A and AC-21-303.1A, and ISO 9001 These differences are readily identifiable, and systems meeting ISO 9000 (ANSI/ASQC91) (*sic)* (*should read ISO 9001(ANSI/ASQC Q91)*) would go a long way toward meeting the FAR Part 21 requirements *but need some enhancements for overall alignment* (emphasis added).

Such misconceptions and partial misrepresentations of the ISO 9000 series are most unfortunate for they indicate a lack of understanding as to what the ISO series is all about. Although one could perhaps speak of appending or complementing such and such standard/regulation to the ISO 9000 series, it is a mistake to speak of "enhancing" or "aligning" the ISO 9000 series with every imaginable standard. Besides the obvious disintegration of the ISO 9000 series which would result from such a process, the approach is in fact contrary to the ISO Technical Committee's Vision 2000 philosophy.

> Vision 2000 emphatically discourages the production of industry/economic-sector-specific generic quality

> standards supplemental to, or derived from, the ISO 9000 series. We believe such proliferation would constrain international trade and impede progress in quality achievements [Donald Marquardt et al., "Vision 2000: The Strategy for the ISO 9000 Series Standards in the '90s," in *Quality Progress*, May 1991, p. 30].

Resistance to the ISO 9000 series has been voiced by other organizations which have perceived the ISO 9000 series as an invasion of their freedom. Still others, mistakenly assuming that ISO 9000 is a *European* Standard, clearly resent having to be told that they should/must adopt a foreign standard! Much of these fears and resistance are, as is so often the case, based on misinformation, a profound lack of understanding as to what the ISO 9001-9003 series of standards is about and an unwillingness to admit that change must be forthcoming.

Which ISO Philosophy Do You Subscribe To?

Having presented well over two dozen ISO 9000 seminars over the past twenty to twenty-four months, I have come across two sets of values/belief/convictions regarding the "correct" philosophical model (i.e., paradigm) for ISO. One faction (which I will label the *pragmatic* group) perceives the ISO series as a set of standards which needs to be addressed sooner or later. This first group is invariably and primarily motivated by a commercial/marketing/business set of convictions. The other faction (the *visionary*) sees the ISO standards as something more than a model for quality assurance. Indeed, this group will more often than not see in ISO an opportunity to improve managerial style. Visionaries speak of ways in which the various ISO models can help a company reduce internal cost and/or increase efficiency. They even speak of the ISO series as an opportunity to develop a Total Quality Management philosophy (some even prefer to refer to the ISO series as a *quality management philosophy* and refuse to use the words quality assurance system).

Who is right? I suppose it all depends on how you envisioned the ISO 9000 series. In my opinion and technically speaking, the *pragmatics* are closer to the truth simply because ISO 9001, 9002 and 9003 are in fact, as the documents' title do indicate, models for quality assurance and NOT models for total quality management. If however one includes the non-contractual standards (particularly 9004/Q94), then indeed many principles of total quality, as preached by Deming, Juran, Crosby, Feigenbaum, Ishikawa and others, are incorporated within the ISO series. However, since 9004/Q94 is nothing more than a set of guidelines NOT audited by any registrars, its paragraphs, valuable and logical as they may be, are nothing more than recommendations.

As the reader is about to find out, the author's bias favors the *pragmatic* faction. This does not in any way suggest that I disagree with any of the propositions put forth by the *visionaries*, on the contrary. I do find it difficult, however, to refer to the 9001-9003 series of standards whose origins can be traced as far back as 1945 when NATO published its NATO 6-49 and the December 1963 Military Specifications MIL-Q-9858A (Quality Program Requirements) and MIL-I-45208A (Inspection System Requirements), as total quality management systems. [Note: Some of the quality assurance systems which have influenced the ISO 9000 series would include the U.S. Air Force Standard AF 5923 which in turn was incorporated within NATO's 1955 AQAPs series, the British DEF STAN 05-21 *et al.*, British Standards 5750 Parts 1, 2 and 3 (see J.M. Juran, "World War II and the Quality Movement," in *Quality Progress*, December, 1991, p.23].

Although I do occasionally refer to the ISO 9004/Q94 document, my primary objective is to guide and inform the reader on how to achieve registration. I do realize that in doing so I may well alienate the *visionaries* who see the ISO series not as a mere registration procedure but as a golden opportunity to implement a long awaited total quality management philosophy. Nonetheless, my experience has demonstrated that companies, or rather individuals, who try to append a total quality "program" on top of an ISO implementation process run a greater risk of stressing the process to the point of break. This is not because the two processes (TQM and ISO) are incompatible but rather because, for a variety of reasons, most individuals or companies do not allocate enough time to carefully plan and smoothly integrate the two processes (see Chapters 3 and 15 for further comments).

The above remarks should not discourage or frustrate the *visionaries*. Indeed, as most companies who have successfully implemented an ISO 9000 quality assurance system have come to find out, when done properly, the ISO implementation efforts do in fact unavoidably lead to a long process of team efforts, shared commitment and improved communication; the very basic foundation of total quality management.

Having reviewed some of the basic issues of the ISO 9000 series, we shall now focus our attention on the 9001/Q91 Standard: the Model for Quality Assurance in Design/Development, Production, Installation, and Servicing.

2 Overview of the ISO 9001/Q91 Requirements

The ISO 9001 (ANSI/ASQC Q91) standard is the most comprehensive of all three models for quality assurance. It consists of twenty paragraphs, some of which are broken down into as many as eight sub-paragraphs (see Table 2.1). Some paragraphs contain an appended NOTE. These NOTES should be viewed as explanatory footnotes and are not part of the standard's requirements.

The primary scope of the 9001 standard is to ensure that a supplier can demonstrate that his quality assurance system is organized in such a way has to prevent the occurrence of nonconformities across all stages from design to servicing. Before proceeding any further, three key terms mentioned throughout the standards should first be defined: supplier, purchaser and sub-contractor.

Unlike other standards which refer to the organization supplying a product or service, i.e. *your organization*, as either the *company* (MIL-Q-9858A and MIL-I-45208A), the *manufacturer* (Good Manufacturing Practices GMP), or the *company* (Malcolm Baldrige Quality Award and others); the ISO 9001-9003 series refers to your company as the *supplier*. Your suppliers are referred to as *sub-contractors* and your customers are called *purchasers*.

The following overview is, for the most part, a re-phrasing of each of the twenty paragraphs comprising the ISO 9001 Standard. Interpretive comments have been kept to a minimum. For further details on how to interpret the standard, the reader is referred to Chapters 3 and 4.

Table 2.1 ISO 9001/Q91 Sub-heading lists

4.1 Management Responsibility

4.1.1 Quality Policy
4.1.2.1 Responsibility and Authority
4.1.2.2 Verification Resources and Personnel
4.1.2.3 Management Representative
4.1.3 Management Review

4.2 Quality System

4.3 Contract Review

4.4 Design Control

4.4.1 General
4.4.2 Design and Development Planning
4.4.2.1 Activity Assignment
4.4.2.2 Organizational and Technical Interfaces
4.4.3 Design Input
4.4.4 Design Output
4.4.5 Design Verification
4.4.6 Design Changes

4.5 Document Control

4.5.1 Document Approval and Issue
4.5.2 Document Changes/Modifications

4.6 Purchasing

4.6.1 General
4.6.2 Assessment of Sub-contractors
4.6.3 Purchasing Data
4.6.4 Verification of Purchased Product

4.7 Purchaser Supplied Product

4.8 Product Identification and Traceability

4.9 Process Control

4.9.1 General
4.9.2 Special Processes

4.10 Inspection and Testing

4.10.1 Receiving Inspection and Testing
4.10.2 In-Process Inspection and Testing
4.10.3 Final Inspection and Testing
4.10.4 Inspection and Test Records

4.11 Inspection, Measuring, and Test Equipment

4.12 Inspection and Test Status

4.13 Control of Nonconforming Product

4.13.1 Nonconforming Review and Disposition

4.14 Corrective Action

4.15 Handling, Storage, Packaging, and Delivery

4.15.1 General
4.15.2 Handling
4.15.3 Storage
4.15.4 Packaging
4.15.5 Delivery

4.16 Quality Records

4.17 Internal Quality Audits

4.18 Training

4.19 Servicing

4.20 Statistical Techniques

The ISO 9001 Model for Quality Assurance in Design/Development, Production, Installation, and Servicing

As specified in its title, the standard is applicable in contractual situations when a supplier must demonstrate his capability to design, develop, produce, install and service a product or service. I should (re)-emphasize that an organization need not satisfy ALL twenty ISO 9001 clauses before being considered for registration for ISO 9001. Indeed, clause 0.0, INTRODUCTION of Q91 does state that the Q91 model for quality assurance is to be used by suppliers when conformance to specified requirements is "to be assured... during several stages which *may include* design/development, production, installation, and servicing" [ANSI/ASQC Q91, p. 1]. Clause 4.19 (Servicing) for example, often does not apply. How many clauses can be "ignored" or labelled as "Not Applicable," is subject to different interpretations (see Chapter 4, p. 59 *et passim*).

A supplier's ability to conform to the ISO 9001 standard is assessed via the standard's *Quality System Requirements*; a set of twenty paragraphs each designed to address a specific portion of a quality system.

4.1 Management Responsibility

This paragraph consists of five sub-paragraphs (see Table 2.1). The *Quality Policy* section was written to ensure that the supplier has defined and documented a written quality policy. The standard goes beyond simply assessing whether or not a policy has been written or duly framed. It also requires the supplier to "ensure that the policy is understood, implemented, and maintained at all levels in the organization."

Under 4.1.2.1, *Responsibility and Authority*, the supplier needs to define the responsibility, authority and interrelation of all personnel whose work affect quality. Since the focus is on who has responsibility and authority to control nonconforming products or services, the supplier has to define (not necessarily by name), which function(s):

- Initiate action to prevent the occurrence of nonconformity.

- Keep records of all identified quality problems.

- Ensures that solutions/corrective actions are implemented.

- Monitor(s) the processing, delivery or installation of nonconforming product until corrective actions have been taken.

Section 4.1.2.2, *Verification Resources and Personnel*, requires the supplier to ensure that verification activities (inspection, test, design verification (see 4.4.5), process monitoring, etc.) are performed by trained personnel (see 4.18). Moreover, and perhaps most importantly, these activities must be carried out "by personnel independent of those having direct responsibility for the work being performed." This ensures that manufacturing functions for example, are not given the responsibility of inspecting their own work (as is sometimes the case!).

Management Representative (paragraph 4.1.2.3) basically states that an appointed management representative shall have the responsibility and authority to ensure that requirements specified in 9001, 9002 or 9003 are implemented and maintained.

The last paragraph, *Management Review* (4.1.3) addresses the issues of timely audits conducted at an "appropriate interval," to ensure suitability and effectiveness of the quality system adopted to satisfy the particular standard (9001, 9002 or 9003). As is the case with most other paragraphs, records must be kept (see paragraph 4.16).

4.2 Quality System

This paragraph requires the supplier to have implemented an effective documented quality system which ensures that the product/service conforms to specified requirements. Some of the components included in the quality system would consist of: documented procedures and instructions (work instruction, test procedures, etc., see other paragraphs), preparation of a quality plan (for new products) and a quality manual, measurement identification and capability, etc. See Chapters 5-7 for further detail.

4.3 Contract Review

This paragraph is often perceived as the "marketing" paragraph. However, since various interfaces between the customer-purchaser and several of the supplier's organizations generally do take place, *Contract Review* usually involves more than just marketing. Basically, four criteria are assessed under this paragraph:

- How does the supplier ensure that contractual requirements are "adequately defined and documented"?

- How are differences between customer requirements and product requirements (e.g., product brief for example), resolved?

- How does the supplier ensure that he has the "capability to meet contractual requirements"?

- Are records kept of all of the above transactions?

4.4 Design Control

Eight sub-paragraphs comprise *Design Control*. The gist of paragraph 4.4 is summarized in the *General* sub-paragraph (4.4.1) which states:

> The supplier shall establish and maintain procedures to control and verify the design of the product in order to ensure that the specified requirements are met (ANSI/ASQC Q94, p. 2).

This involves ALL design activities including:

- How are design plans drawn up and updated during the development cycle? This include an explanation of who is responsible for what activity (4.4.2).

- What are the technical interfaces between the various design groups? What type of documented information flows across these various interfaces and are they regularly reviewed? (4.4.2.2, see also 4.4.2.1 *Activity Assignment* regarding the use of qualified staff and "adequate resources.)

- Who is responsible for identifying, documenting, verifying and resolving ambiguous or conflicting input design requirements (4.4.3)?

- How are design output requirements documented and how do they meet input requirements, conformance to regulatory requirements, as well as other safety and functional requirements (4.4.4)?

- What competent personnel is/are assigned the task of verifying the design (4.4.5)?

- What procedure(s) is/are in place to identify, document and periodically review "all changes and modifications" (4.4.6)?

4.5 Document Control

The *Document Control* paragraph consists of two sub-paragraphs: 4.5.1 *Document Approval and Issue* and 4.5.2 *Document Changes Modifications.* These paragraphs are often found to be in noncompliance. Sub-paragraph 4.5.1 states that document control procedures as they relate to the requirements of ISO 9001 (9002 or 9003), shall be established. The important issues to consider while addressing this paragraph are:

- What control procedures are in place?

- Who reviews and approves documents?

- How does the supplier ensure that appropriate documents "are available at all locations where operations essential to the effective functioning of the quality system are performed"?

- What procedure is in place to ensure that obsolete documents are "promptly" removed from all locations?

Paragraph 4.5.2 focuses on procedures relating to document changes and modifications. These changes should, unless otherwise specified, be performed by the issuing functions/organizations. Related issues would include:

- How are changes identified within the document?

- How and where are current revisions identified?

- How many revisions before a document/section is re-issued?

4.6 Purchasing

This paragraph sets requirements relating to the *Assessment of Sub-Contractors* (4.6.2), *Purchasing Data* (4.6.3) and *Verification of Purchased Product* (4.6.4). The opening sub-paragraph (4.6.1) of section 4.6 sets the tone for the remaining three sub-paragraphs by stating that "The supplier shall ensure that purchased product conforms to specified requirements." In order to do so, the standard informs us (4.6.2) that the supplier shall base the selection of his sub-contractors based on their ability to meet sub-contract requirements. The selection process and type of quality assurance control used to assess sub-contractors shall also depend on the type of product as well as history of past performance. Records (a variety of forms and formats are acceptable) of "acceptable" sub-contractors and some measure of the quality system's effectiveness also need to be considered.

Sub-paragraph 4.6.3 deals strictly with the accuracy, adequacy, identification, specification and other documentation review procedures relating to purchasing documents (approval or product qualification procedures, drawing specifications, etc.).

Paragraph 4.6.4 addresses special situations when the purchaser (i.e., your customer) reserves the right, as per contractual requirements, to verify "at source or upon receipt that purchased product conforms to specified requirements." Such, at the source verifications, by the purchaser do not absolve the supplier of all responsibilities relating to product acceptability. Nor should it be interpreted as evidence of effective "quality control by the sub-contractor."

4.7 Purchaser Supplied Product

This paragraph addresses issues relating to products (e.g., sub-components), supplied by the purchaser to the supplier for assembly or incorporation by the supplier. In such cases, the supplier must have control procedures, including record keeping, relating (as applicable) to verification, storage, and maintenance of the purchaser's supplied product.

4.8 Product Identification and Traceability

This is one of the few "where appropriate" paragraphs (see Chapter 3 for further references on how to read the standard). Interpretation of this and other similar paragraphs depends on the type of industry you are in. As indicated in the title, this paragraph addresses product identification and traceability during all stages of production, delivery, and installation. Naturally, the paragraph's significance is relative to the type of product being manufactured. The pharmaceutical, medical, food and related industries do devote much more attention to this paragraph than for example, a manufacturer of foam used for packaging.

4.9 Process Control

This is yet another of those "where applicable" paragraphs. The opening sentence compounds the confusion and does not facilitate interpretation. Indeed, paragraph 4.9.1 opens by stating:

> The supplier shall identify and plan the production and, where applicable, installation processes *which directly affect quality* and shall ensure that these processes are carried out under controlled conditions (ANSI/ASQC Q91, p. 4).

What *directly affects quality* is not always easily determined. Some would say everything, others would say very little. We will return to these issues in Chapter 9. Suffice it to say for now that, once you have determined what directly affects quality, you will have to ensure that these activities are carried under controlled conditions. This means that you will have to:

- Document work instructions which relate to production and installation. To help us focus on how this can be done, the standard goes on to state that documented work instructions are needed "*where the absence of such instructions would adversely affect quality*"! If your customers have not already done so, it is up to you to define what would adversely affect quality.

- Monitor, control and approve processes during production and/or installation.

- Determine criteria of workmanship using various means.

Special Processes (4.9.2) was written to cover processes whose effectiveness can only be measured after the product is in use. The classification of processes as "special" and "standard" is not as easy as it might first appear. In some industries, engineers would have you believe that most of their processes could be classified as "special." One must be careful with such statements because as the standard explains, special processes are processes "the result of which cannot be fully verified by subsequent inspection and testing of the product" or where deficiencies can only be detected after the product is in use. As one seminar participant once noted, "I guess air bags would be classified as special products!" I hope not.

4.10 Inspection and Testing

Receiving Inspection and Testing (4.10.1), *In-Process Inspection and Testing* (4.10.2), *Final Inspection and Testing* (4.10.3) and *Inspection and Test Records* (4.10.4) are the four sub-paragraphs of 4.10. There are basically two major points to consider when reading the two sub-paragraphs of 4.10.1 *Receiving Inspection and Testing*:

- Incoming product shall be inspected or otherwise verified for specified requirements as specified in the supplier's quality plan.

- The release of incoming product for urgent production purposes shall be done according to specific procedures which allows for the immediate recall and replacement "in the event of nonconformance."

In *In-Process Inspection and Testing* (4.10.2), the supplier must ensure that a set of documented procedures, as specified in the quality plan, are in place to:

- Inspect, test, and identify product.

- Ensure that the product conforms to specified requirements.

- Guarantee that a product is not released until "required inspection and tests have been completed." (See also 4.10.1.2.)

- Identify any nonconforming product.

Paragraph 4.10.3 *Final Inspection and Testing* basically covers the same principles as those outlined in 4.10.1 and 4.10.2 except that it focuses on final inspection and testing of the finished product. The product cannot be dispatched until all inspection activities have been satisfactorily completed and documented. The last paragraph (4.10.4) simply states that records of all inspection and test results shall be maintained.

4.11 Inspection, Measuring, and Test Equipment

This is the longest paragraph. It consists of ten sub-paragraphs (a-j). The emphasis of 4.11 is on: calibration, measurement accuracy and precision for inspection, measuring and test equipment. For all of these activities, records of inspections (including frequency of inspection) will have to be kept. Instruments will have to be tagged or otherwise identified to indicate last and next calibration date. Other related issues concern the validity of previous inspection and test results, the suitability of environmental conditions for testing procedures and the safeguarding against "adjustment which would invalidate the calibration setting." See the ANSI/ASQC Q91 Standard for further details.

4.12 Inspection and Test Status

Throughout production and installation, the inspection and test status of the product must be identified using "suitable means." Moreover, whenever nonconforming product is released, records will need to be established in order to identify the inspection authority responsible for the release.

4.13 Control of Nonconforming Product

The standard's emphasis on the control and disposition of nonconforming product is clearly reflected in this paragraph as well as 4.13.1 *Nonconforming Review and Disposition.* The following set of questions should be asked with regards to 4.13 and 4.13.1:

- What procedures are in place to prevent the inadvertent use or installation of nonconforming product?

- How is the nonconforming product identified? Does it need to be segregated? What organization(s) need to be notified?

- Who is responsible for the disposition (rework, re-grade, reject or accept as is) of nonconforming?

- Are you contractually required to report and record concessions to the customer? If so, who has the responsibility and authority?

For further details, see ANSI/ASQC Q91 and subsequent chapters.

4.14 Corrective Action

Corrective actions consists of five sub-paragraphs (a-e). These paragraphs were included to ensure that the supplier has developed procedures designed to:

- *Investigate* the cause(s) of nonconformity and implement effective and lasting corrective actions which will *prevent recurrence*!

- Analyze all relevant quality records including service reports, customer complaints, work instructions, process controls and concessions "to detect and *eliminate potential causes of nonconforming product.*"

- Initiate preventive actions commensurate with the potential risk.

- Verify that corrective actions have indeed been implemented.

- Ensure that the necessary procedural changes have been implemented and recorded as a result of said corrective action(s).

4.15 Handling, Storage, Packaging, and Delivery

The four topics of *Handling* (4.15.2), *Storage* (4.15.3), *Packaging* (4.15.4) *and Delivery* (4.15.5) are included to ensure that the supplier has documented procedures which guarantee proper handling, secure storage designed to prevent damage or deterioration of product, packaging which conforms to specified requirements and delivery procedures which maintains the integrity of the product after final inspection and test.

The importance and procedural detail of each of the above paragraphs will naturally depend on the type of industry. For some companies, *Storage* is not an issue simply because the product's half life far exceeds any specification (assuming such specifications do exist). For some companies (manufacturer of crystal vases, for example), *Packaging* is a most crucial procedure.

4.16 Quality Records

This paragraph is cited in twelve of the twenty paragraphs. Obviously, it is an important paragraph. Its focus is essentially on quality record keeping, maintenance, retention, storage and availability.

4.17 Internal Quality Audits

Once a quality system is in place it must be audited to ensure that it is effective and indeed complies with said requirements. This is achieved via the mechanism of *internal audits*. These audits must be planned, scheduled and (of course) records of findings and follow-up actions must be maintained. In addition, "management personnel responsible for the area shall take timely corrective action on the deficiencies found by the audit" (see 4.1.3).

4.18 Training

Training is the most costly activity carried on by a supplier. The training needs for all personnel whose work affects quality must be identified by the supplier. For example, internal auditors will have to be trained prior to conducting internal audits. As in most other paragraphs, records of all training shall have to be maintained. Training generally goes beyond "quality related" type activities and includes safety and technical training.

4.19 Servicing

This paragraph applies only to companies who are contractually obliged to provide servicing activities. For some companies, paragraph 4.19 is Not Applicable. For others, for whom servicing is an important activity, paragraph 4.19 merely states that servicing procedures shall meet specified requirements! Defining what are the specified requirements can be a most interesting exercise for companies who do provide servicing. The author has participated in some meetings where, after a couple of hours of intense discussion, the participants finally agreed that a considerable amount of effort would be required to address paragraph 4.19 (see also clause 16.2 of Q94).

4.20 Statistical Techniques

This last paragraph is often misinterpreted as requiring the supplier to implement Statistical Process Control (SPC) techniques. Such is not the intent of paragraph 4.20. This four line paragraph begins by stating that "where appropriate," the supplier shall identify "adequate statistical techniques for verifying the acceptability of process capability and product characteristics." No reference is ever made to SPC or other techniques. The standard assumes that the reader will know what is meant by "process capability." However, this might be assuming too much since it is not clear whether or not the author(s) of paragraph 4.20 had in mind statistical process capability (Cpk ratio for example). The only reference to specific statistical techniques can be found in paragraph 20.2 *Statistical Techniques* of ISO 9004/Q94. Sub-paragraph (e) mentions "quality control charts/cusum techniques" (ANSI/ASQC Q94, p. 18).

In most cases, your customers will more than likely specify which statistical techniques they would like you to use. Invariably, SPC is one of those techniques.

Conclusions

The above review of the twenty paragraphs of ISO 9001/Q91 should give the reader a good overview of the standard. The review is not intended to be a substitute for the actual perusal of ISO 9001/Q91. Readers who are seriously considering implementing an ISO 9001, 9002 or 9003 quality assurance system must purchase the appropriate standards (see Appendix B for a list of addresses and phone numbers).

This chapter's primary objective was to present each paragraph and offer a few interpretive comments to help the reader better assess what is required of the standard. Some companies/individuals (including the author) have relied on the ISO 9004/Q94 *Guidelines* to interpret ISO 9001, 2 or 3. There is nothing wrong in doing so as long as one realizes that the paragraphs of the ISO 9004/Q94 *Guidelines* are NOT requirements. This trite observation might appear to be redundant but the fact is that too many people begin reading the *Guidelines* and, forgetting that they are *Guidelines*, make erroneous assertions as to what the ISO standard(s) require. The reference to Statistical Process Control (paragraph 4.20), is but one such example. The ISO 9004/Q94 standard is an excellent document which everyone should read. It may well be that in the years to come updates to the ISO 9001-3 standards will come from the ISO 9004/Q94 document [some have even suggested that with some modifications, ISO 9004 should eventually replace 9001]. However, the truth of the matter is that, as of now and for the next five years, the only requirements (i.e., requirements subject to third party audits), will come from either

9001, 9002 or 9003. Having reviewed the ISO 9001 standard let us now interpret its jargon.

3 How to Interpret and Address ISO 9001

I have selected the ISO 9001/Q91 document simply because it is the most comprehensive of all three standards. In most cases, the phraseology between standards is nearly identical, the only difference (to be corrected in future editions of the standards) is in the numbering of the paragraphs.

Satisfying each of ISO 9001, 9002 or 9003 sub-paragraphs, as the case may be, involves a substantial amount of preparation. As has already been demonstrated, individuals who will have to adapt their quality system to one of the three ISO models may have to consider addressing as many as twenty major headings (for ISO 9001/Q91), *many of which are interrelated.* This interrelationship between paragraphs is all too often ignored in favor of a modular approach which unfortunately does little to help break down existing departmental fortresses.

Anyone involved with the implementation and/or documentation of an ISO 9001, 2 or 3 quality assurance system is impressed by the emphasis on documentation and record keeping. This is hardly surprising since as Table 3.1 and Figure 3.1 indicate, paragraph 4.16, Quality Records (4.15 in Q92), is the paragraph most often cross-referenced. [Note: All paragraph numbers refer to the ANSI/ASQC Q91 document.]

Record keeping is certainly an important aspect of the standards however it does not necessarily lead to a nightmarish orgy of record maintenance. In most cases, companies already have good to fair record keeping. What needs to be done is to ensure, among other things, that the location of all records is identified, that pertinent records are up to date, reviewed as need be, as well as properly archived and retained for a pre-determined period of time.

Table 3.1: Cross-Reference of ISO 9001 Paragraphs

4.16 Quality Records	4.1.3 Management Review 4.2 (g) Quality System 4.3 (c) Contract Review 4.4.5 (a) Design Verification 4.6.2 Assessment of Sub-contractors 4.7 Purchaser Supplied Product 4.10.1.2 Receiving Inspection+Testing 4.10.4 Inspection and Test Records 4.11(f) Inspection,Measuring+Testing 4.12 Inspection and Test Status 4.13.1 Nonconforming Review 4.18 Training
4.18 Training	4.1.2.2 Verification Resources and Personnel
4.17 Internal Q Audits	4.1.3 Management Review
4.1.3 Mng. Review	4.17 Internal Quality Audits
4.4.4 Design Output	4.4.5 Design Verification

How to Read the Standard

Most people, including consultants, tend to interpret the standard much "too rigidly." As already stated in the previous chapter, this is in part due to the fact that some people rely on the ISO 9004 standard to interpret 9001, 9002 or 9003. Within the context of 9001, 9002 and 9003 phraseology, there are in fact very distinct grades of requirements.

- The **shall** sentences

These sentences are found throughout the document. Paragraph 4.1.1 for example begins with: "The supplier's management **shall define and document**..." The majority of ISO paragraphs have a "**shall**"

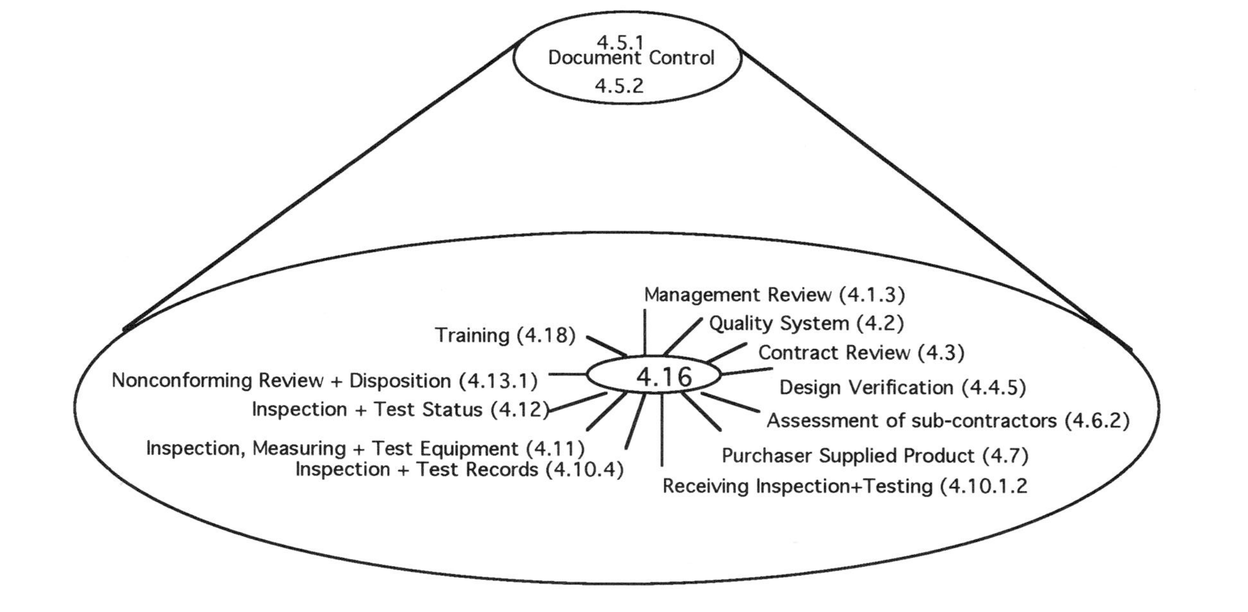

Figure 3.1 Interrelation Between Paragraphs

clause, fortunately (or unfortunately, depending on your point of view), none of the three standards specify *how you shall do it*, only that you must do it! It is therefore up to **you** to decide *how you will implement the requirement* (assuming you are not currently satisfying it).

- Other less restrictive sentences

These sentences would include statements such as:

- *"Where practicable..."* (4.5.2, 4.6.3, etc.)
- *"A master list or equivalent document control..."* (4.5.2)
- *"...where appropriate, on records of subcontractors."* (4.6.2, 4.8)
- *"...or other suitable means."* (4.12)

- Other key sentences:

- *"...where the absence of such instructions would affect quality,..."* (4.9.1 a)
- *"...until it has been inspected or otherwise verified as conforming to specific requirements."* (4.10.1.1)
- *"inspect, test, and identify product as required by the quality plan or documented procedures;"* (4.10.2)
- *"The supplier shall control, calibrate, and maintain inspection...to demonstrate the conformance of product to the specified requirements."* (4.11)

Additional Comments Regarding Interpretation

The difficulty in implementing an ISO 9001, 2 or 3 type quality assurance system rests with the interpretation of its many sub-paragraphs. When first reading the standard, people usually have difficulty interpreting its content. The paucity of precise instructions on what to do and how to do it frustrates anyone who has to translate the generic sentence structure of the standard into a quality assurance system. For example, under **Statistical Techniques**, one reads the following:

> *Where appropriate, the supplier shall establish procedures for identifying adequate statistical techniques required for verifying the acceptability of process capability and product characteristics* (ANSI/ASQC Q91-1987, paragraph 4.20, p. 7).

The decision to use "adequate statistical techniques" is left for client/supplier interpretation. This flexibility of interpretation is both a source of strength and weakness of ISO phraseology. Its strength is that unlike other supplier evaluation schemes, the standards do not impose a particular procedure and/or technique on a supplier. Because the document is intended to be a model or a set of *guidelines* covering a broad range of industries, its authors had to purposefully *suggest* all encompassing guidelines which would be interpreted by each industry as they saw fit. The unavoidable difficulty with guidelines is that there are as many interpretations as there are readers and assessors.

Let us focus on the first paragraph of ISO 9001, paragraph 4.1.1, Quality Policy.

> *The supplier's management* ***shall*** *define and document its policy and objectives for, and commitment to, quality. The supplier* ***shall ensure*** *that this policy is* ***understood, implemented, and maintained at all levels*** *in the organization* (ANSI/ASQC Q91-1987, p. 1).

Most companies address the issue of quality policy by issuing a one page statement supposedly written and signed by the highest ranking official. The policy is invariably framed and strategically placed in various offices. Is that all that is required? Certainly not! What is most important as far as the standard is concerned is that the policy is *understood, implemented and maintained at all levels in the organization.* The most important issue then, is not necessarily to have a policy but to *ensure* (perhaps via training) that the policy is practiced by everyone.

How to Address a Paragraph: An Example

For many companies, sub-paragraphs 4.5.1 and 4.5.2 of *Document Control* are often a source of confusion. The most often asked question regarding paragraph 4.5 is: "How do I do it?" What are the important points of sub-paragraphs 4.5.1 and 4.5.2?

4.5.1 Document Approval and Issue

> *The supplier* ***shall establish and maintain*** *procedures to control all documents and data that relate to the requirements of this Standard. These documents* ***shall be reviewed and approved*** *for adequacy by authorized personnel prior to issue. This control shall* ***ensure*** *that:*
> *a) the pertinent issues of appropriate documents are available at all locations where* ***operations essential to the effective functioning of the quality system are performed****;*
> *b) obsolete documents are promptly removed from all points of issue or use* (ANSI/ASQC Standard Q91-1987, p. 3).

All too often, quality managers or consultants alike, simply rephrase or slightly edit the above paragraph and include it in the company's quality manual. The problem with such an approach is that few, if any, employees are interviewed to verify if indeed the affected departments or functions do indeed follow the stated procedures! If you should decide to somewhat arbitrarily write procedures, you should not be surprised to find out that on the day of the audit, third party assessors will have no difficulties raising nonconformities (see Chapter 12: The Third Party Audit).

The reader must realize that he/she is not a servant of the standard. The strategy is NOT to write lengthy and detailed procedures which you think will please the assessor's interpretation of the standard, but rather to adapt your current procedures to fit the requirements as stated in each of the standard's paragraphs. Most companies already have hundreds of procedures, unfortunately they are invariably *out of date and therefore inaccurate.*

The easiest method to follow is to simply rephrase the paragraph into a series of questions. For example, with regard to paragraph 4.5.1, one could begin by asking:

> "Do I (we) have procedures to control documents and data?" [If the answer is "No," you will either have to write some or come up with a pretty good reason to explain why not.]
>
> "How are documents/procedures approved? Who approves procedural changes?" "Where can I find out?" As you begin your search, you will often find out that procedures do exist BUT that no one follows them! [Another good reason to be issued a nonconformance.]

As you begin improving on your quality assurance system the temptation will be to include more and more layers of detailed instructions. If such a temptation should occur, remember to read and re-read paragraph *4.5.2 Document Changes/Modifications*. The more detailed your procedures, the more likely they will be affected by changes and the more often they will have to be revised, updated, re-written, signed, approved, etc! *Therefore, keep it simple and concise.* [See Chapter 5 for some examples of how to control documents.]

Word Processors, Electronic Mail and Document Control

Today's word processors can ease the task of controlling your documents. Most software allow you to print on each page of your documents, footers, headers, date and even time of day. In addition the computer's own data management will keep track of when files were saved and last accessed. Computerized data management does not resolve all issues however. Someone must still decide, perhaps with the help of the Management Information Systems department, who will approve the document? Who and how the documents will be controlled — what hierarchy of passwords and super passwords will be needed? How and to whom they will be distributed? In some organizations or departments where personal computers are abundant, electronic mail (E-mail) can ease much of the burden of document control, modification and distribution.

A Word of Caution Regarding the Use of Word Processing

In an effort to reduce paperwork a pharmaceutical company decided to "word process" all of its laboratory procedures. Having stored all procedures onto the centralized computer, authorized parties could now easily update the centralized procedure files and quickly inform all interested parties via electronic mail. Everyone seemed to approve of the new computerized procedure except the purchasing manager. Someone had failed to recognize that all customers had to be informed of process changes (particularly lab procedures). The purchasing manager was faced with the enormous task of having to inform every customer, via official correspondence, of each and every change and/or modification! Paperwork which had been eliminated by one department had simply been transferred to another.

Conclusions

Having reviewed how the standards need to be interpreted, suggestions were offered on how to answer some specific requirements. The most direct approach is to simply rephrase each paragraph into a series of questions. The reformulation of the ISO 9001/Q91 standard into a questionnaire is presented in Appendix C. You may use the questionnaire to do a preliminary assessment of your company's readiness towards the 9001 quality assurance model (for 9002 simply ignore section 4.4).

Before embarking on the documentation effort, key questions must first be answered, these include: Which standard is most appropriate for your needs? Which segment of the organization will have to be registered? How will you proceed and organize your efforts to guarantee success?

4 The ISO Quality Assurance System: How to Proceed?

> *The quality system of an organization is influenced by the objectives of the organization, by the product or service, and by practices specific to the organization, and therefore, the quality system varies from one organization to another* [ANSI/ASQC Standard Q90-1987, p. 1.]

Prerequisites to the Quality Assurance System

Whatever the definition of quality (and there have been many), one must first focus on three fundamental elements: *Quality Policy, Quality Management* and *Quality System* in order to achieve quality.

Quality Policy

Overall quality objectives and direction of an organization pertaining to key elements of quality such as fitness for use, performance, safety and reliability, as formally expressed by top management.

Quality Management

That aspect of the overall management function that establishes and implements the quality policy.

Quality System

The organizational structure, responsibilities, procedures, processes, and resources used for implementing quality management.

(Adapted with some rephrasing from ANSI/ASQC Standard Q90-1987, pp. 2-3.)

In order to successfully implement a quality system, organizations must recognize that the above "qualities" (policy, management and system), are symbiotically linked to each other in a logical hierarchical network. Most implementation errors are generally attributable to quality policies which either refute the importance of the executive's commitment to a *Quality Policy*, downplay the need for management's engagement towards the stated policy and/or hastily organize and adopt a *Quality System* without the knowledge, support or consent of those directly affected by it.

The ISO Quality Assurance Model and Total Quality Management

There tends to be some confusion or occasional philosophical disagreements when someone tries to compare the ISO 9001, 2 or 3 series of quality assurance standards with the principles of Total Quality Management (TQM). When people realize that a significant portion of the implementation efforts required to achieve registration consist of documenting what you say you do, some individuals develop an allergic reaction to ISO. To them, the real issue is total quality management. To achieve that objective everyone must be involved to solve a multitude of problems which may have accumulated over several years. Some even go as far as suggesting, with some sarcasm, that registration to one of the ISO standards "doesn't even guarantee that we will make good products!"

Such criticisms are not well founded for several reasons. First of all, as has already been stated (Chapter 1), the ISO series of standards is a model for the management of a quality assurance system designed to insure that at a *minimum*, a series of steps are taken to ensure that you do indeed satisfy your customer requirements. Whereas it is true that the ISO 9001-3 series will never question your technical ability to satisfy your customer specifications, some of its paragraphs have specifically been written to verify that you have an effective system in place to do so. What confuses some individuals is that strong directives on how to achieve these goals are not specifically written in the standards. Perhaps this is for the better (see Chapter 15 for additional comments).

As most businesses begin to implement a quality assurance system they quickly notice that a vast network of internal customer/supplier relationships begin to develop. When done properly and in a spirit of cooperation, the setting-up of a ISO 900?/Q9? quality assurance system can lead to a sharing and exchange of information across previously impermeable departmental barriers. The phenomenon is similar to the Total Quality Loop schematically depicted on page four of ANSI/ASQC Standard Q94-1987 (see Figures 4.1 and 4.2).

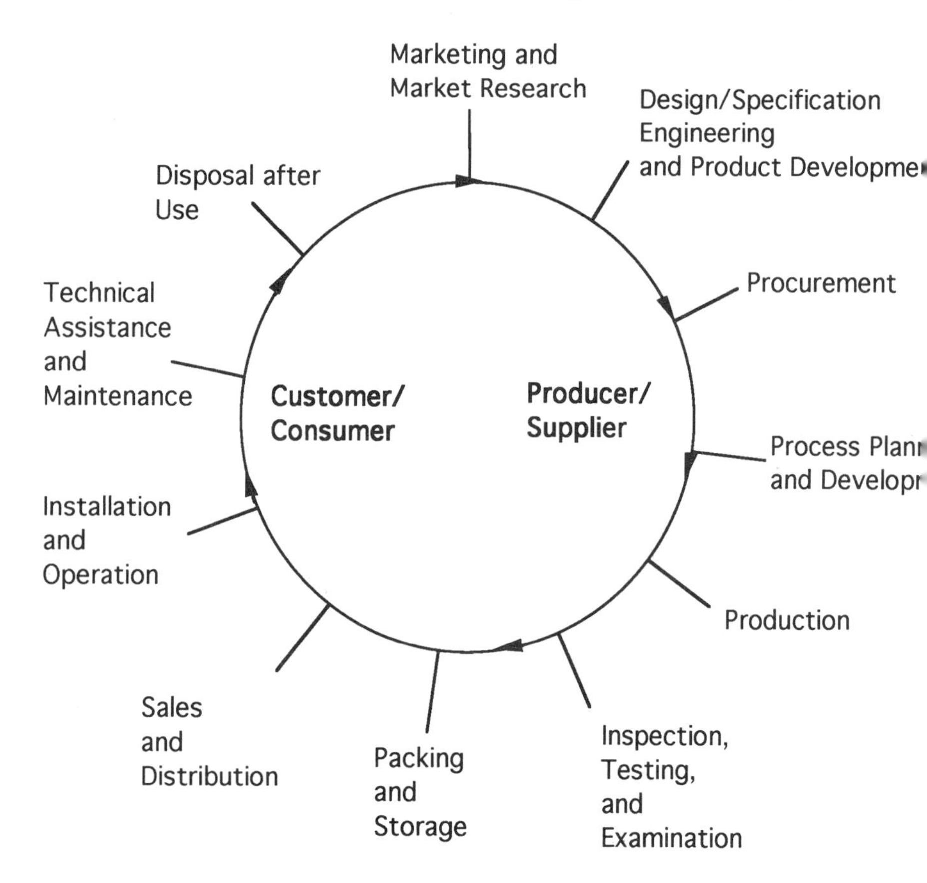

Figure 4.1 The Quality Loop (adapted from ISO 9004)

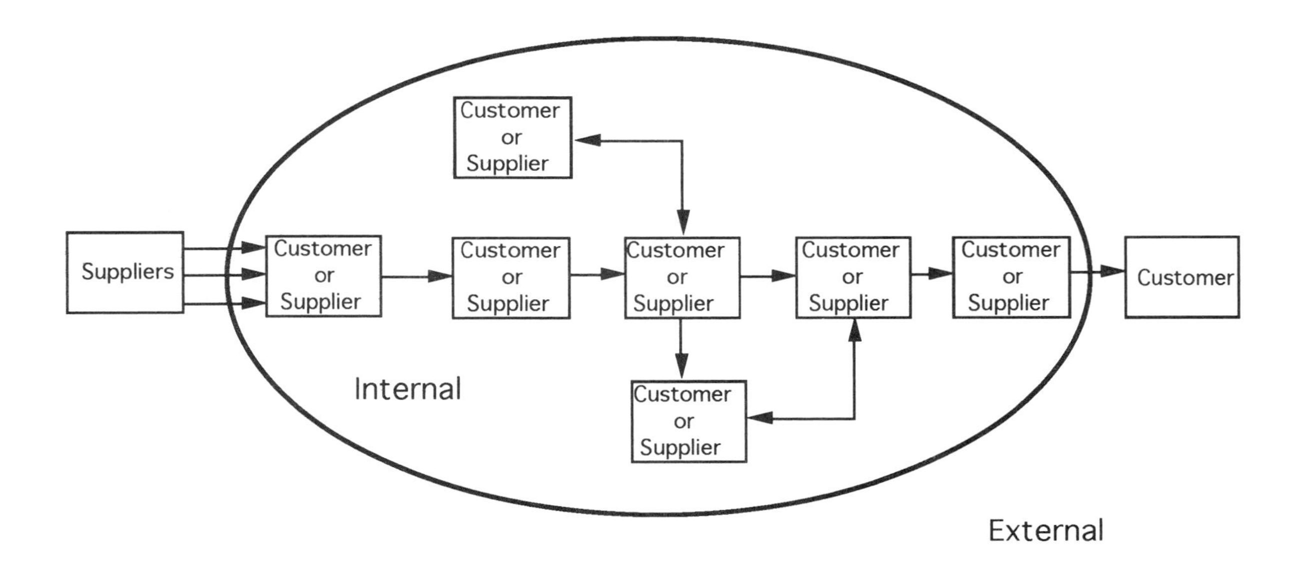

Figure 4.2 Customer Interaction

Scope and Field Application: Issues to Consider

Despite the many seminars offered to help alleviate the "fear" of ISO 9000, there is still much misunderstanding as to what "ISO 9000 requires." The fact is that you are not really required to do things to "please" ISO, or its interpreters (the assessors). Rather, most companies or individuals who have expressed an interest in the ISO series have done so because their customers have requested, required, demanded or otherwise stated that ISO certification will be a requirement in the near future. But what do the standards state with regard to "the customer"? ISO 9002 (ANSI/ASQC Q-92), for example, explains under paragraph **1.0 Scope and Field of Application**:

> **1.1 Scope**
>
> This Standard specifies quality system requirements for use **where a contract between two parties** requires the demonstration of a supplier's capability to control the processes that determine the acceptability of product supplied.
>
> The requirements specified in this Standard are aimed primarily at preventing and at detecting any nonconformity during production and installation, and implementing the means to prevent recurrence [ANSI/ASQC Q92-1987, p. 1].

In essence then, the standards are nothing more than a set of requirements which should *already be in place* (for the most part) whenever two parties enter into some sort of contractual agreement. In fact the standards do recognize that although "these Standards will normally be adopted in their present form,... *they may need to be tailored for specific contractual situations."*

Thus, before undertaking your documentation efforts, you should first ask yourself a very basic question:

- What is the nature of our contractual agreements regarding product quality? Or, if we are to use the words of the ISO 9002 standards for example, how do you ensure or otherwise demonstrate to your clients that the control you exercise over your processes guarantees the delivery of an acceptable product?

Although the answer to that question is probably "it depends on the client/product," there must be a basic contractual philosophy which applies to the majority of your clients. It is important to assess what your contractual philosophy is because it will help you determine how to answer some of the questions you might have regarding ISO. Consequently, I would advise that as soon as possible and with the help of those involved with contract review, your first step should be to flow chart the contractual process or processes (see Chapter 8: Documenting Your Procedures). Doing so will prove most helpful in the later stages of the implementation plan.

How to Proceed?

My experience with companies which have adopted one of the three ISO quality assurance model, is that they tend to form into two groups: those that seem to act as though they have NO or very little documentation, and those that would have you believe that everything is in order. In either case, the truth is often somewhere in between. If it were indeed true that no or little documentation were available, then certainly much would have to be written. But even if that were the case, one should not focus on how much will need to be written but rather on what will have to be written and how it can help a plant improve its performance by reducing its cost of quality. Moreover, it is very likely that many (verbal) procedures and documentation do already exist — hence the optimism of companies which belong to the second group. Of course, as is often the case, most of what is available (in written form) is probably out of date, hence the need to establish a document control procedure! Still, one of the first tasks to accomplish during the implementation procedure would be to collect and validate all available documents. Once all available documentation has been collected, identified or otherwise sorted, the next step would be to decide to which tier (two or three, see Chapters 5 and 6) it belongs and to determine which ISO paragraph it addresses.

If your organization maintains its quality system, you may very well find out that as much as three quarters of the work is already done. Companies with much experience with the Department of Defense, the Aeronautical/Aerospace or akin industries, or companies currently having to satisfy their customers' Supplier Quality Assurance (SQA) systems could, in some cases, have as much as seventy (or more) percent of their current quality system satisfy the ISO Standards.

Organizational Issues and Certification

Before looking into as to how the quality assurance system should be organized, you must first ask yourself some important questions:

- Why ISO certification (purpose, objectives)?
- Which ISO (9001 or 9002)?
- Who or what is driving the effort?
- Which product line(s) (all, a few)?
- Which plant(s) or portion thereof will be certified?
- If plants or functions are geographically separated, how will the task be coordinated?
- How to proceed? One pilot plant followed by more plants or all plants simultaneously?
- Any internal sub-contractors?
- Will external sub-contractors have to be certified?
- Available resources (one man or one committee)?
- What is the timetable?
- What level of commitment and from whom?
- Will there be a need to coordinate efforts with other

committees (OSHA for example)?

The organizational configurations of most companies are often not well adapted to the implicit structure outlined within the ISO quality systems. The boundaries between process ownership and documentation control for example, are not always clearly delineated nor do they always follow a linear flow as is often the case in assembly line type manufacturing. For example, Figure 4.3 represents a typical (although much simplified) chemical plant.

- Who is/are the suppliers?
- Which standard would apply, 9001, 9002, 9003 or all of them?
- Where are the product or plant boundaries?
- Who or what will be certified?
- Should R&D be treated separately? How about purchasing?

Fortunately, as has already been stated, the ISO/TC 176 committee does recognize that "on occasions they (Standards) may need to be tailored for specific contractual situations" [ANSI/ASQC Standard Q92-1987, p. 1].

Case Studies Relating to "What to Register?"

Case Study 1.

A manufacturer of electronic components is geographically and functionally organized as follows:

- Research and Development ========> United States
- Manufacturing of sub-components ===> South Korea
- Some manufacturing + final assembly=> France

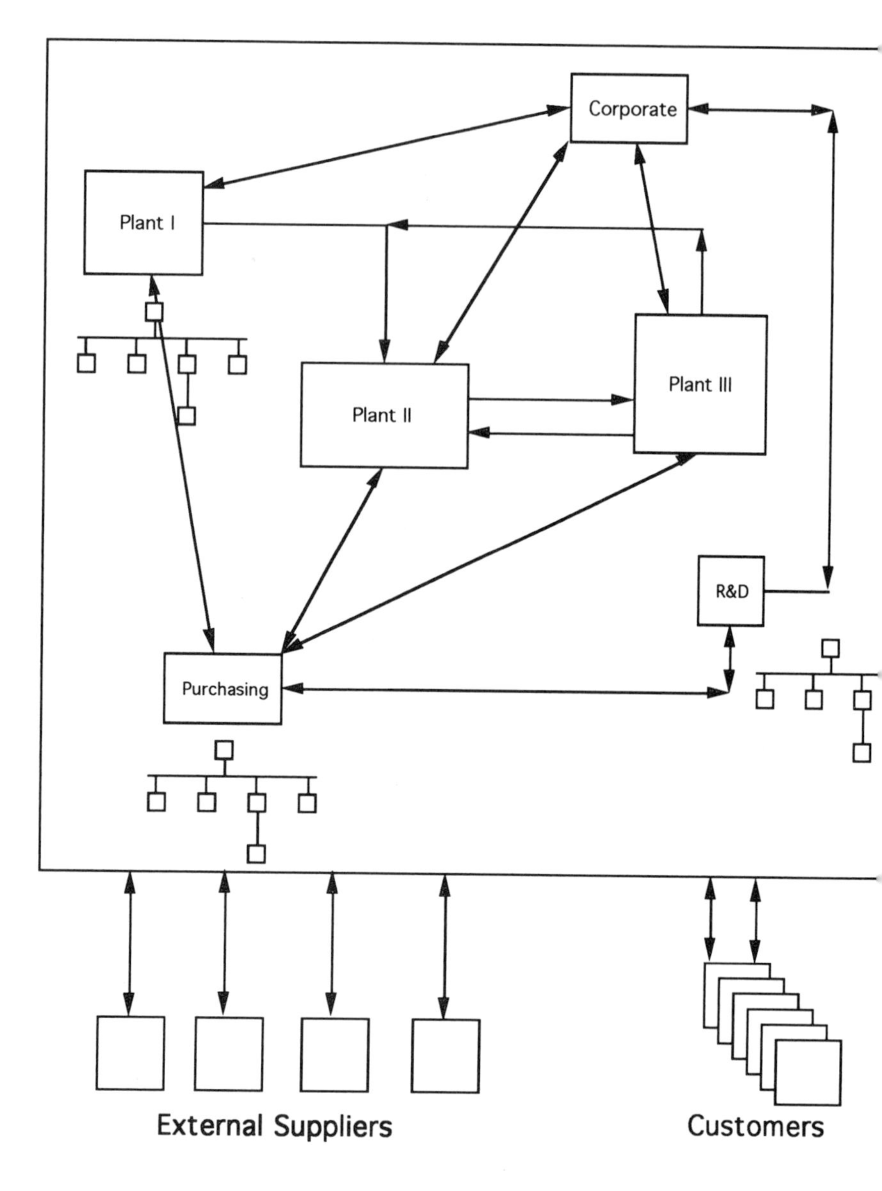

Figure 4.3 What would you register?

Research and Development is jealously guarded by the U.S. The French plant (which has some R&D activity) receives many of its sub-components from South Korea which manufactures according to U.S. designs. How and what would you register?

Initially the U.S. plant wanted to "go for ISO 9001." This seemed a logical approach since R&D was involved. However, after asking a few questions and explaining that exchange of information would have to ensue between the various sites, I was quickly informed of the impracticality of such an idea simply because the U.S. R&D team was not about to exchange information with France, let alone South Korea! Such xenophobic behaviors are not at all unusual even among the most seasoned multinational corporations. Having worked in Europe with a few U.S. and foreign multinationals, I can testify that although English is the international language of business and engineering, cultural biases and prejudices are still deeply rooted on both sides of the Atlantic. Still, how should this particular multinational proceed?

When confronted with the above scenario (which is rather typical), managers often ask me: "What must I do to satisfy ISO?" The answer to that question cannot be found in any of the ISO 9001, 9002 or 9003 paragraphs. Nor can it be addressed by any of the registrars. Nonetheless, when pressed for an answer, registrars or consultants will try to come up with an answer. Some registrars seem to come up with some strange convoluted interpretation as to what can be registered to the ISO 9000 series. One participant once explained to me that a major registrar recently (last quarter of 1991) registered the R&D division of a major company to ISO 9001. Perplexed by his comments, I asked how the registrar managed to apply the twenty clauses of ISO 9001 to an R&D division? "They apparently dismissed non-applicable clauses and only retained the relevant ones," was his answer! Fair enough, I suppose, but if that is indeed how the registrar proceeded, one would like to know how many clauses were dismissed as irrelevant. Was 4.9 dismissed? What constitutes nonconforming product within an R&D environment (4.13)? My experience years ago as an assistant chemist in food R&D taught me that most "product" in applied R&D IS nonconforming. Corrective actions (4.14) consisted of yet another experimental run. I suppose one can indeed justify the registration of an R&D group to ISO 9001, although I still question the practice. If some registrars do indeed register individual units within an organization, one can anticipate the following scenario: Company XYZ has its R&D registered to ISO 9001, its manufacturing facility registered to ISO 9002, its laboratory and warehouse registered to ISO 9003, its purchasing registered to ISO 9002 (why not?), its quality department registered to ISO 9003, etc. Moreover, isn't it slightly misleading to say that a company is registered to ISO 9001 when in actuality it is only the R&D division which is registered to ISO 9001?

In the case of the above multinational, the preliminary (and unsatisfactory) recommendation was to have the South Korean plant apply for ISO 9002 and the French plant apply for ISO 9001. If and when the U.S. were to join in the ISO registration efforts, U.S. registration to ISO 9001 could eventually be considered — in fact, if we are to accept the approach explained in the above paragraph — the U.S. R&D could even independently apply for ISO 9001. These decisions must be reached by top management. If the current *modus operandi* of the U.S. multinational is perfectly acceptable to the president and vice presidents of the various divisions, the ISO series of standards cannot offer advice, nor is it empowered to do so.

Case Study 2

This is a particularly unusual, however not necessarily uncommon, scenario. This particular company operates as follows:

- Internal suppliers consisting of two divisions located in one building: (a) R&D (family run) facility which formulates highly secretive formulas and (b) a group of "mixers" who combine ingredients according to specifications passed on by the R&D group. Note that in this particular instant, the "mixers" are never told what the chemical formulas are; they are simply given mixing ratios and procedural steps describing in what order the ingredients should be mixed.

- Once mixed, bags of mixed components are sent to an internal customer (manufacturing) who in turn combines the ingredient with components supplied by external suppliers.

- The Quality Control group can only verify *finished* product quality according to some pre-determined (by the R&D group) product characteristics. When nonconformities are observed, the Q.C. group can only report the nonconformance but cannot take any action. All concessions are made by R&D which strangely enough does not request Q.C. reports.

Could such a company register for ISO 9001? Probably not. The best it could achieve would be to try to register its manufacturing facility to ISO 9002. Unless R&D is willing to cooperate, it would be difficult to document a quality assurance system which would adequately address the issues of nonconformance as specified in any of the ISO standards.

This reluctance on the part of R&D to participate in any "ISO implementation efforts" is rather common. I have experienced many such cases in the United Kingdom, France, Germany and naturally, the United States. This is not to say that R&D will never cooperate, but rather that R&D staff generally tend to be the last to join in the efforts. The most common argument (or shall we say excuse), generally put forth by scientists is that the R&D process, being a creative process, cannot be controlled via the type of documentation required by the ISO series of standards. I will let the reader reach his/her own conclusions as to what the real reason(s) might be.

Other case studies could be presented but they tend to be variations of the same theme. When considering registration to one of the ISO standards, an organization should consider the issues raised in the above case studies as well as ask itself the following questions:

(1) Do we want to register only a product or line of products?

(2) Should we register one or more plants?

(3) Should we consider a corporate wide registration scheme? If we do what are the issues to consider and how shall we approach the task?

The first approach (product registration), is the least effective and, as some would argue, perhaps the least rational — think of the difficulties and cost involved when one tries to distinguish and differentiate between ISO vs non-ISO products. Irrespective of which approach you choose, unless your customer dictates a particular approach, *you and only you can decide how to proceed.*

Conclusions

Having noted the importance of the symbiotic relation between management's commitment to a policy of quality and the success of the implementation efforts, the chapter concludes with an analysis of the important questions to ask when considering registration. On the assumption that you will not begin your implementation efforts without first attempting to ask all the right questions and answer them to the best of your ability, we can now proceed to the next step: how can you organize your quality assurance model to facilitate your registration procedure?

5 Some Suggestions on How to Organize a Quality Assurance System: The Pyramid of Quality Model

Preliminary Comments

Several options are available when documenting a quality assurance system. As far as the ISO 9000/Q90 standard is concerned, the quality system should be "documented and demonstrable in a manner consistent with the requirements of the selected model" (ANSI/ASQC Q90-1987, paragraph 8.3, p. 4). Demonstration means that the elements of the quality system are not only adequate but also ensure that product or service conformity comply to specified requirements. Documentation, we are told, "*may* include quality manuals, descriptions of quality-related procedures, quality system auditing reports, and other quality records." (ANSI/ASQC Q90-1987, p. 5). (For further information, particularly as it relates to "the nature and degree of demonstration," the reader is referred to the Q90 standard.)

The model herewith presented should not be construed as being *required* by any of the ISO standards. It is in fact a generic approach which has evolved thoughout the years thanks to the efforts of a multitude of individuals. The model happens to be favored by most registrars who in turn have borrowed it from various sources. I should emphasize that when it comes to interpreting the ISO standards, three factions are involved:

- The potential customer of ISO (i.e., you the reader).

- The registrars, or third party accredited body, which will assess your organization through its own interpretation of the standard and,

- The standards themselves, written by the TC 176 committee.

The potential customer can purchase and read the standard. However the customer cannot seek advice from members of the TC 176 committee, unless he/she happens to know one such member. But even then, since registration to any of the three ISO standards is not issued by the TC 176 committee but rather by the *interpreters* of the standards, namely, the registrars, one needs to understand how the registrars generally perceive the standards (see Chapters 11 and 12 for further information on registrars). Hence the emphasis and bias in this and the next two chapters on the registrar point of view.

Although no requirements specify or even outline how the quality system should be organized, many "experts" and reputable registrars seem to have been influenced by paragraph 5.3.2 entitled *The Quality Manual* of ISO 9004 *Quality Management and Quality Systems Elements — Guidelines.* Referring to 5.3.2 of ISO9004/Q94, and more specifically to 5.3.2.4, one reads that for large companies, the "quality management system may take various forms," including a corporate, divisional and specialized quality manuals. Based on that suggestion, some of the leading registrars have recommended, and indeed strongly suggested, that *one* efficient way to organize a quality system would be to adopt the three tier approach known to all who have attended one of the many "official" Lead Assessor Training courses as the *Quality Pyramid.* This structure (see Figure 5.1) recognizes three types of documents: Tier one (the quality manual), tier two (departmental procedures) and tier three (work instructions). We will first focus on tier one documentation.

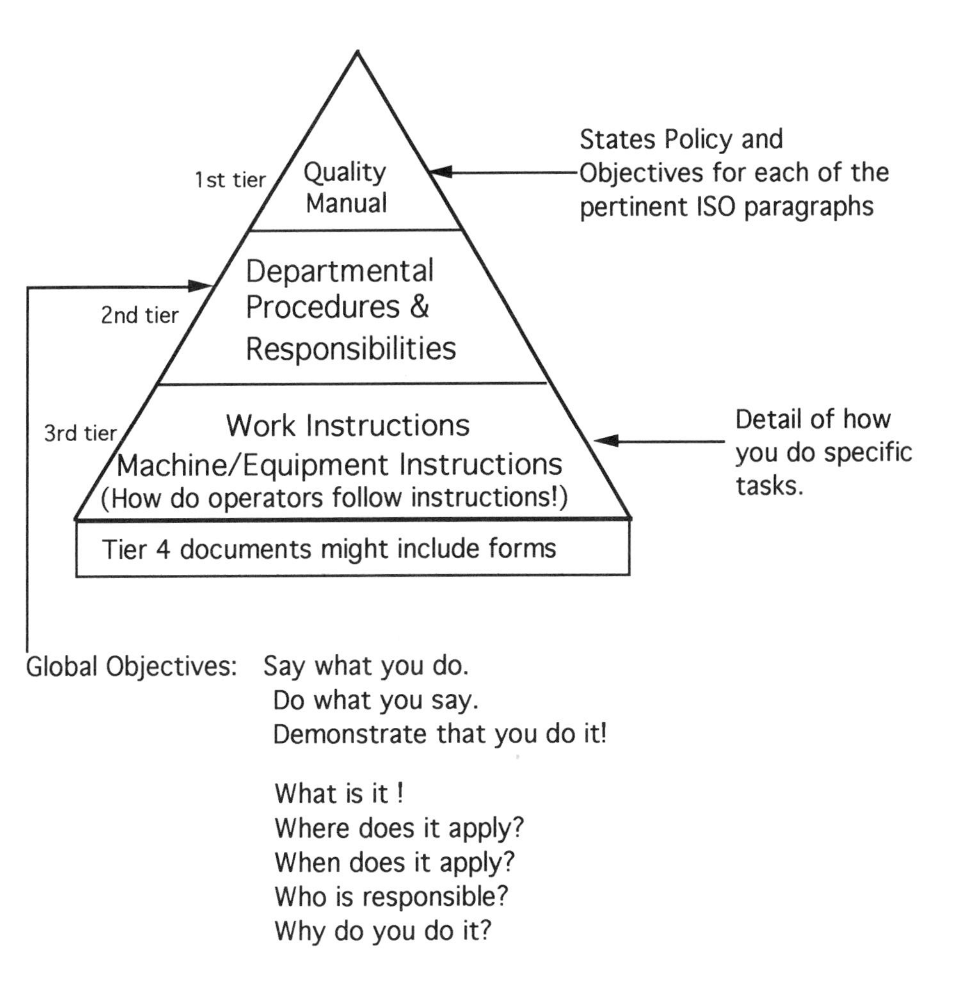

Figure 5.1 The Quality Pyramid

Tier One: The Quality Manual

Although the Q90 standard states that a quality manual *may be included*, the truth of the matter is that, to the author's knowledge, most if not all registrars do require that a quality manual be written and submitted.

Consultants and auditors alike do not always agree as to what constitutes the model quality manual. Although most would agree that each of the ISO (9001 or 9002) paragraphs *should* be addressed in the quality manual (even if a particular section is non-applicable), different opinions are voiced when it comes to style, content, page format, control policies, and other related issues.

The quality manual is the top part of the pyramid. As a document, it should be brief and concise, not exceeding 25-35 pages. Contrary to traditional quality practices in this and other countries, the detailed documentation of procedures and work instructions should **NOT** be part of the quality manual. Rather, such procedures should be bound in the appropriate tier two or tier three documents.

The quality manual must address each of the sections of either ISO 9001, 9002 or 9003. It need not be detailed in content. Its purpose is to assure the reader/assessor that the company in question has indeed addressed all of the relevant ISO paragraphs. How each department addresses each task is detailed in tier two documentation which in turn might refer to tier three work instructions or Standard Operating Procedures. Appendix A provides some guidelines on what needs to be included in a quality manual.

Format of the Quality Manual

There is no set style on how to write a quality manual. Irrespective of the style, the quality manual should have a particular format. The format suggested herewith (see following pages and Appendix A) is generally accepted by the leading registrars.

The format should include the name of the company or plant on the top of each page (or at least the first page) as well as have the "document control box" at the bottom or top of each page.

Following the title page (see example on page 67) and table of contents, one of the first pages should be the *Revision List* page. Various formats are available for the revison page, the example shown below is but one such format.

Date	Page	Paragraph #	Comment	Approval: Name and Signature.

The *Distribution List* page can follow the *Revision List* page. Its format can vary.

Department/Office Location	Person Responsible for Manual
Note: Manual may be borrowed or reviewed by anyone and thus may not always reside permanently at said location. Person in charge of department is however responsible for manual at his/her location.	

A common mistake is to include too much detailed information. Examples of information that should *NOT* be inserted in the quality manual would include:

- Laboratory procedures.

- Technical specifications.

- Names of individuals responsible for specific tasks. For example, including the name of managers is not advised for the simple reason that every time there is a name change, an updated version of the quality manual (a controlled document) would have to be issued.

It is important to remember that the quality manual is a *controlled document* and that as such, a control procedure must be specified. Some of the important questions to ask while preparing the quality manual are:

- Who should receive the manual (distribution list)?

- How will revisions be handled and by whom?

- How will the document be controlled?

- How will you ensure that obsolete quality manuals are removed from circulation? Who will be responsible?

- Who will address the "Management Responsibility" section?

- Who will write the quality manual?

- How will the task of coordinating all the paragraphs be addressed?

- How will you ensure that no unreasonable and paradoxical claims are made throughout the manual or across tiers?

- Who will review the manual for accuracy prior to final release? I would suggest everyone listed on the Distribution List.

- What should you include in the control box?

After you have considered and addressed all of the above questions, you may be as creative as you wish. See Appendix A for a more detailed model of how to organize and what to include in a quality manual. Some quality manuals (particularly those written by consultants or agencies) can be so generic that they can apply to any industry. In fact, some agencies even have quality manuals already pre-formatted and ready to be modified on their word processor to suit any client's need. I would not be surprised if in the months to come a company were to market a "quality manual software package" where the user will only have to input a few key words, style preference, select one of many scenarios and the software will take care of the rest: format, document control, etc.

A typical quality manual format, consists of only three sections for each of the pertinent ISO clauses. Essentially duplicating the format and structure of the ISO document themselves, the sections are: 1.0, *Scope*, 2.0, *Policy and* 3.0, *Responsibilities.* Under *Scope* the manual's author basically states the purpose of the section. The *Policy* section describes in very generic terms, usually broken down into several sub-paragraphs (2.1, 2.2, etc.), the actual company policy regarding the particular ISO clause. The *Responsibility* section explains who (without naming anyone) is responsible for the policy stated in the above paragraph(s). These quality manuals often read as a mere rephrasing of the standard (see next page). This is not to imply that their style should be avoided. In fact, many U.S. multinationals who have subsidiaries in the U.K. tend to adopt, perhaps too readily and sometimes with only minor modifications, the quality manual of their British facilities.

The JLQC Company	Revision: 1.1	Page 1 of 1
	Last Update: 10/90	Date: 11/15/91
Title: CONTRACT REVIEW	Title: Quality Manual	Section No. 3.0
This page is copyrighted	Approved: J.L.L.	Written by: S.H.L.

1 SCOPE

This section defines the contract review of the JLQC company.

2 POLICY

2.1 All contracts are reviewed according to documented procedures.

2.2 Documents are reviewed for accuracy. Inaccuracies or conflicts are reviewed and resolved with the customer.

2.3 Nonconformities are documented and analyzed by the department.

3 RESPONSIBILITIES

3.1 The review of all tenders is the direct responsibility of the staff assigned the particular account.

3.2 The manager is responsible for resolving inaccuracies or conflicts.

3.3 The department is periodically audited for system effectiveness and improvement opportunity by functions independent of purchasing.

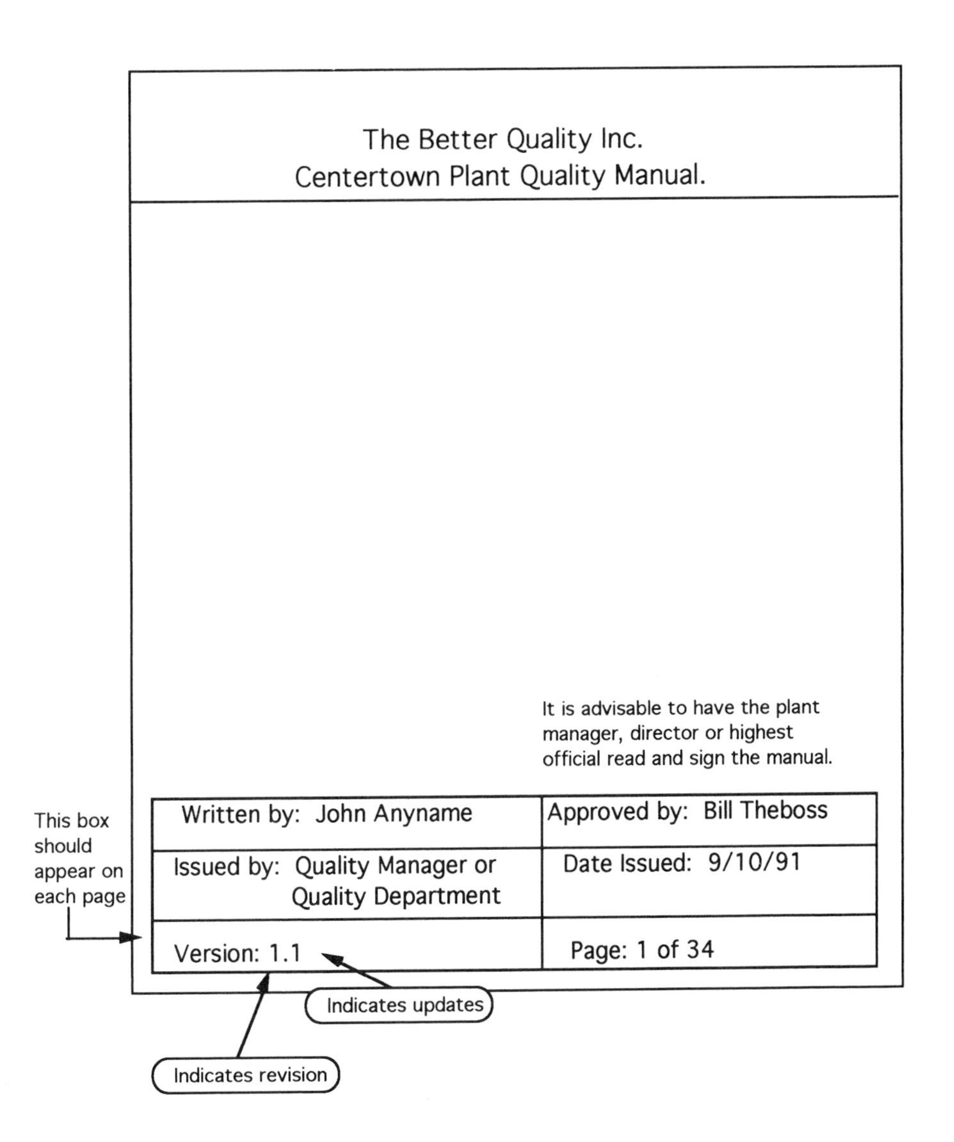

The format "Version 1.1" indicates that every 10 updates a complete updated revision is re-issued. This is required by the document control sections of ISO.

Conclusion

We have seen that there are literally as many styles as there are writers. Quality manuals which have been written by persons other than "an insider" tend to be more generic than those which have been written by a staff member. Sub-contracted manuals tend to all look alike. One could say that they are almost "too perfect." The document control section is very detailed with a long list of document numbers, procedure or standard operating procedures codes, detailed and often elaborate policy numbers. The style and format of these documents is reminiscent of the DOD or the nuclear industry. Naturally, there is absolutely nothing wrong with these documents except perhaps if they were intended to be for a manufacturer of ball point springs. In such cases, one tends to become suspicious as to who has ownership of the document.

Some companies distribute their quality manual on a write protect diskette. One company in the Midwest exchanges quality manuals with its suppliers by simply mailing diskettes. Whenever a manual needs to be updated, the supplier in question recalls his write protect diskette and sends a new updated diskette. Revisions are maintained on a separate file. Although the process is very effective, I do not know if registrars are, shall we say, sufficiently up to date to accept diskettes from potential clients. Registrars would only have to have access to three or four of the most popular word processors to make the process work.

Many companies seem to want to copyright their manuals; some even insist on including a copyright statement on every page. Although I do understand the need to protect certain sensitive documents, I do not quite understand why an ISO 9000 type quality manual would have to be copyrighted. The information contained in the manual should be made available to all potential customers. In addition, the manual's contents should be generic enough not to reveal any sensitive information. If sensitive information is to be included, it probably should be inserted in tier two or tier three documents.

6 Tier Two Documentation

The writing of tier two documentation is *not* a required activity. Nonetheless, a description of departmental responsibilities can be most beneficial, particularly if internal customers/suppliers are given the opportunity to review each other's perceived responsibilities. Some companies include tier two documentation within the quality manual. Others develop separate tier two (departmental) documentation. The usual procedure used to document tier two documentation is to have each department generate its own set of documents. These documents should be more detailed than the tier one document in that they should:

- Explain the department's functions (**What**).

- May include a departmental organization chart and briefly describe everyone's responsibility, authority, functions/role (**Who**).

- May refer to the appropriate **S**tandard **O**perating **P**rocedures (i.e., tier three documents which explain **how** things are done).

- May state who are the department's internal and/or external suppliers and customers. This is an optional but very valuable step which can easily be summarized in one diagram.

Others prefer to organize tier two documents by simply going back to each of the ISO paragraphs referred to in tier one, and explain in more detail *who* controls processes, *who* is responsible for instrument calibration, *who* performs final product inspection, etc. Although this approach might be practical for small plants, it becomes difficult to manage with large plants, particularly when several organizations are responsible for the same or similar activities.

Finally, one must recognize that some procedures apply to more than one division or department. Document control and in some cases instrument calibration (i.e. metrology), are two such procedures. Regarding document control, there is sometimes no need to have each department have its own unique set of procedures. A corporate or plant policy on document control could be established or might already exist. Similarly for metrology. Some plants have as many as four or five organizations perform calibration depending on the type of instruments. Rather than have each organization write its own second tier documentation, a common document could be produced. Such duplication of efforts is not only redundant and thus wasteful but also not required.

[Note: Some industries prefer to have each department document their "respective" tier two ISO paragraph (see Figures 6.1 and 6.2.). This approach is just as logical *as long as duplication of efforts is avoided*. Indeed, when faced with the task of documenting tier two, the affected department invariably re-generates a complete quality manual to address each of the twenty (9001) or eighteen (9002) ISO paragraphs.]

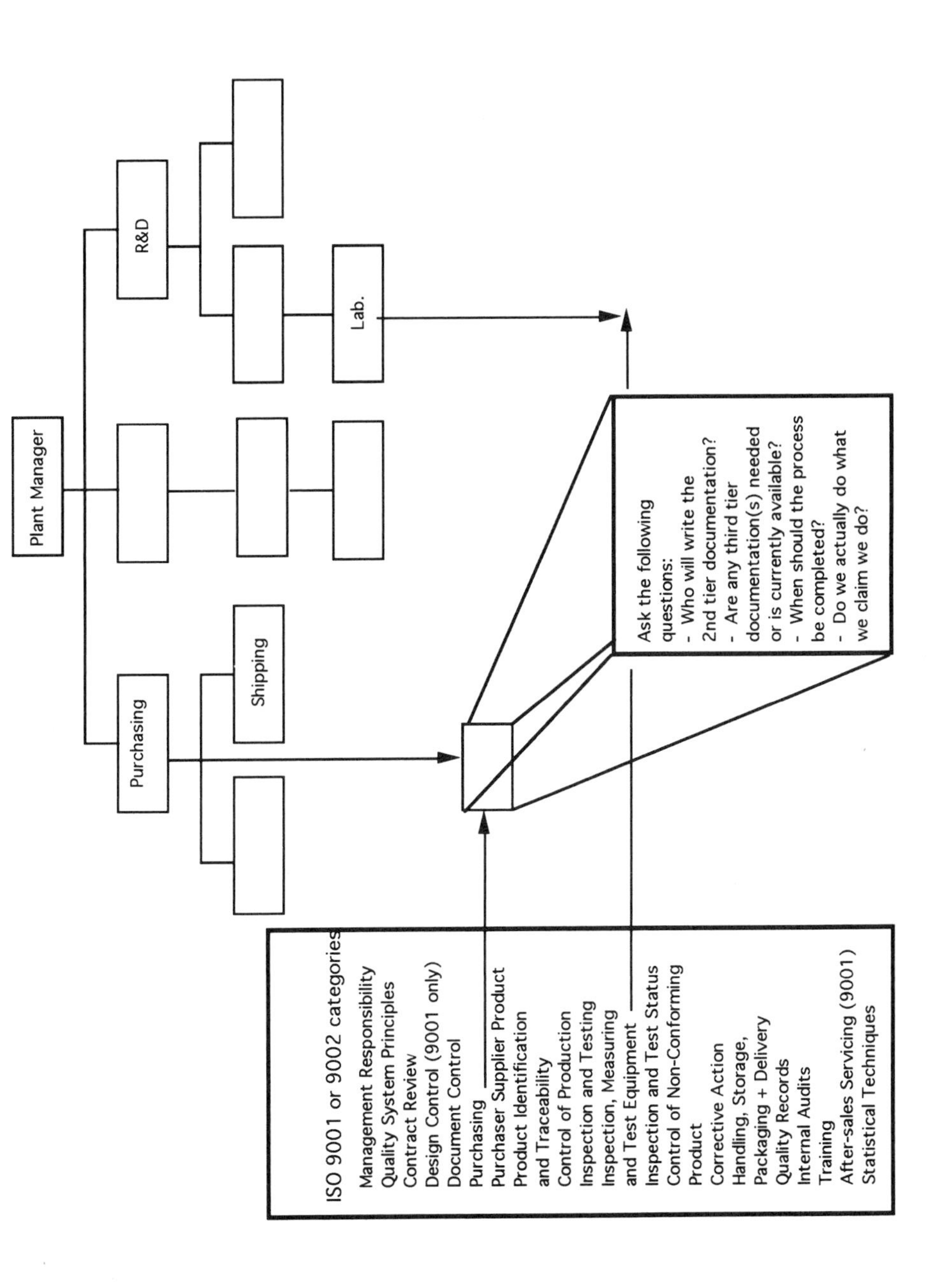

Figure 6.1. Implementation Matrix

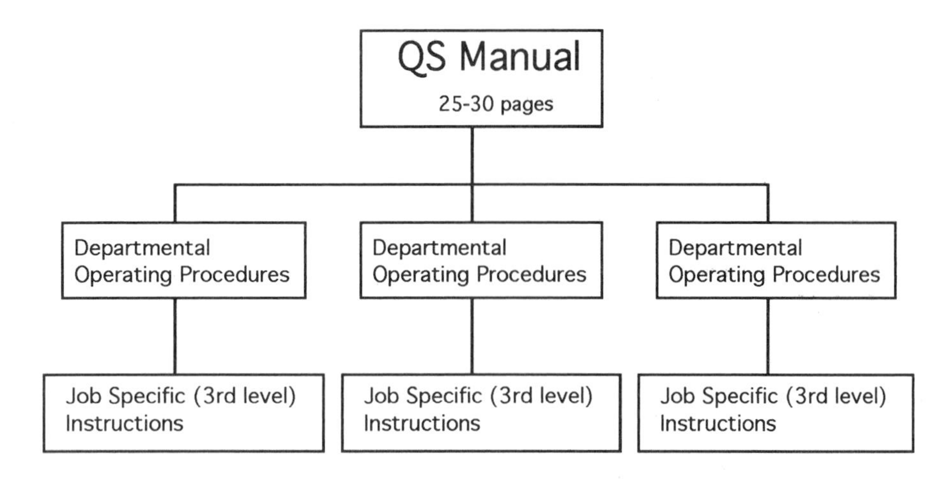

Figure 6.2 Another Way of Organizing Tier Two

The following examples are provided as guidelines and are included only to assist the reader. The formats described in the next few pages and chapters should not necessarily be considered as officially approved formats. They are, nonetheless, good examples of how to document your quality system and are accepted by many registrars. You are encouraged to innovate or otherwise improve and adapt these examples to best suit your needs.

Tier 2 Sample: Method 1

[Note: The following style is rather typical of individuals who write documents by strictly adhering to the content of paragraphs 4.5.1 and 4.5.2 of Document Control without questioning the feasibility of implementing such procedures. There is of course nothing wrong with the content, however, an auditor would like to know if everyone involved with document control is aware of the procedure and its many steps! What other comments positive or otherwise would you make? SOP stands for Standard Operating Procedures.]

Document Control

1.0 Purpose

Provide system of control to insure that all documents and data releases pertinent for the effective operation of the company's quality system are reviewed and approved by authorized personnel prior to issue.

2.0 Scope

Applies to all documents with a critical bearing on the quality of the final product.

3.0 References (optional)

ISO 9002/Q92:1987 Sections 4.4, 4.4.1, 4.4.2
Plant's quality manual.
Unit/Departmental manuals.

4.0 Definitions (optional)

None.

5.0 Responsibilities and Authority

5.1 Document control procedures for the quality manual are specified in the manual.

5.2 Tier two document control consists of the following format:

Document #	Document's Title:
Revision #	Pagination by section #
Prepared by:	Approved by:

The above format is inserted at the top of the first page of each document. Tier two documents are bound in one tier two manual available from the quality manager.

A master list of all documents is maintained by the quality manager. The list indicates:

• Title, name with signature of who updates and has authority over each procedure.

• Where - building and office location — the master procedure(s) are located.

• Revision status of each section for tier one and two documents and by procedure # for tier three documents.

6.0 Approvals

6.1 Document approval, issuance and modification have been decentralized and are the responsibilities of each functional area. Anyone can initiate document revision or new document preparation. Document revision or preparation is initiated by a memo stating:

- The name of the person recommending the action.
- The date the recommendation is submitted.
- Description of recommended revision or new document needed.

The memo is sent to either the person currently filling the position of the last author, or the affected area management for a new document.

6.2 Review and Approval

The proposed change/document is reviewed by area users and experts who are required to provide written or verbal responses. Once all appropriate draft modifications have been made, the final draft is sent to the Department Manager or designee for final review. The Department Manager/designee is responsible for assuring the appropriate input and approval has been attained and that related modifications to associated documentation have been made. If all of these requirements have been satisfied, the document is then issued.

6.3 Issue

Prior to issuing a document, the author must prepare the following:

- Document distribution list.
- Reason for revision/new issue.
- Description of revision/new issue.
- Author signature and function.
- Approval signature and function.
- Date of issue.
- A new table of contents with the correct revision number for each section.

A copy of the Table of Contents is forwarded to the ISO coordinator to update the master list.

Tier 2 Sample: Method 2

0.0 Shipping Section	
Document Title: Tier 2 Shipping	Description: Tier 2 document for Shipping Section
Revision number: 1.0	Page 1 of 4
Prepared by: Mark Swann	Approved by: Gene Forester
Issued by: Shipping Section	Issue Date: June 3 1991

1.0 Purpose

To describe the organizational structure, operation and responsibilities of the Shipping Section.

2.0 Scope

Applies, but is not limited to all activities affecting the overall quality of the final product.

3.0 Departmental Responsibilities

- Receiving of Incoming Raw Materials.
  - Raw material coordination and accounting.
  - Distribution and accounting of catalysts and consumable chemicals.
- Railroad Car Maintenance and Repair.
  - Tank car repair, cleaning and coordination.
  - Certification of tank cars and tank truck scales.
- Interface with Shipping Regulatory Agencies.
- Production Accounting and Planning.
- Delivery of finished product.
  - Preparation of all shipping documentation.
  - Loading of product shipments.

1.0 Management Responsibility

Document Title: Tier 2 Shipping	Description: Tier 2 document for Shipping Section
Revision number: 1.0	Page 2 of 4
Prepared by: Mark Swann	Approved by: Gene Forester
Issued by: Shipping Section	Issue Date: June 3 1991

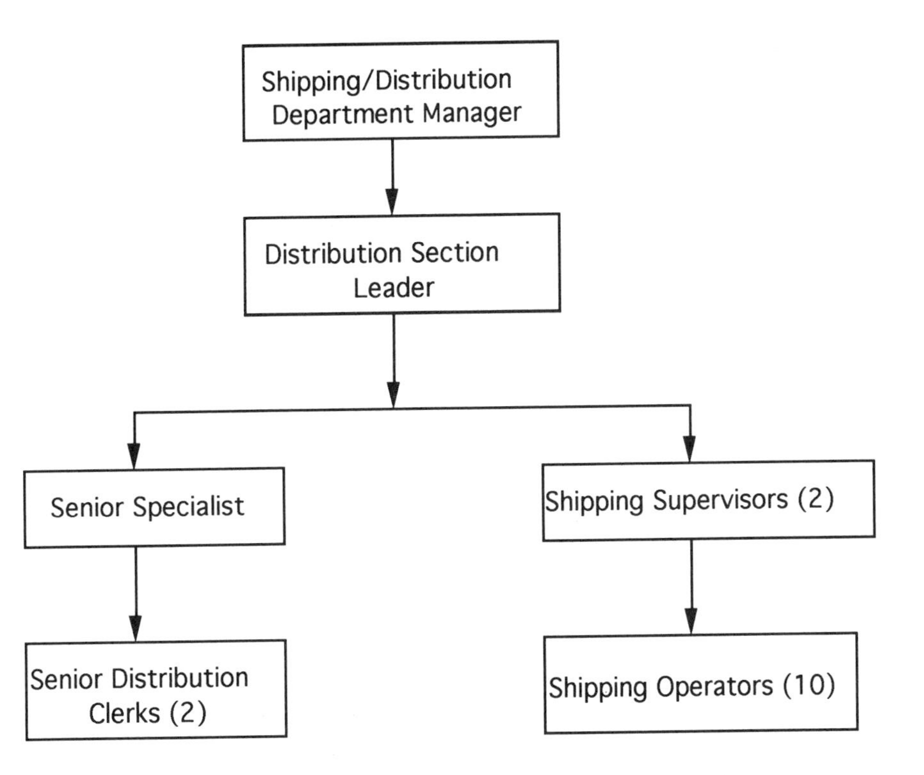

4.0 Individual Responsibilities

1. Department Manager

    - Oversees all aspects of the daily activities listed in 0.0.

2. Section Leader

    - Coordinates all activities relating to incoming raw materials, regulatory compliance, shipment documentation, ownership of hazardous material pipeline, certification of tank cars as well as acts as the plant's Emergency Response Coordinator.

3. Senior Specialist

    - Acts as a liaison between Plant Operation, Management, Hydrocarbons Group and Production Services Supervisors.

    - Determines product availability, sets priorities and orders specialized equipment.

    - Confirms books for each unit's inputs.

    - Reconciles the final Good Book.

4. Senior Distribution Clerk

    - Maintains a plant accountability of raw material usage (fuel gas, oxygen, nitrogen, ethylene and propylene) received from various suppliers.

    - Sends ledger of flows, pressures and temperatures to Accounting Department (SOP #).

    - Prepares a Tank Car Inventory Report.

    - Reviews incoming order and files order by product name in an Open Order Registrar.

    - Prepares billing information worksheet.

Document Title: Tier 2 Shipping	Description: Tier 2 document for Shipping Section
Revision number: 1.0	Page 4 of 4
Prepared by: Mark Swann	Approved by: Gene Forester
Issued by: Shipping Section	Issue Date: June 3 1991

- Collects loading information sheets.
- Distributes Bill of Lading.
- Maintains an Inventory Sheet.
- Maintains a Shipped Report.
- Distributes the Inventory Sheet and Shipped Report to the various Control Rooms.

5. Shipping Supervisor

- Prepares Loading Instruction Sheets.
- Oversees loading activities.

6. Shipping/Loading Operators

- Move available car to wash and repair rack (Ref. S.O.P #).
- Move car to loading spot.
- Pre-load inspection prior to loading (Ref. S.O.P #).
- Weigh cars.
- Take samples (SOP#).
- Prepare Consolidated Loading and Unloading Reports.

[***Note:*** Flow charts (next page) can be a very effective tool to document and update procedures (see also Chapter 8: Documenting your Procedures).]

Incoming Order Procedure

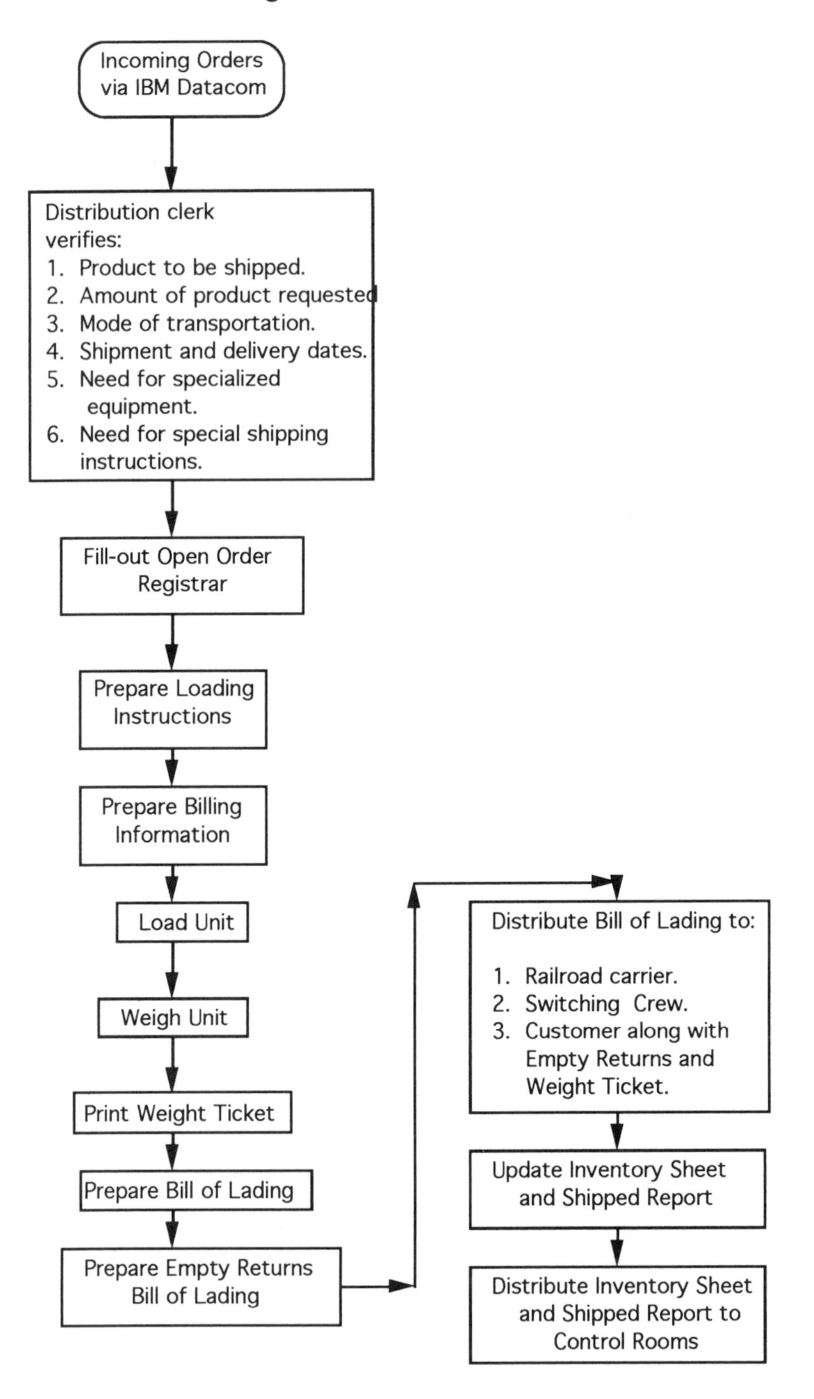

Conclusions

The above examples are but suggestions on how to approach tier two documentation. Many more formats and styles are available. It is perhaps fortunate that, as of yet, no standards are currently available on how to write procedures. This might change in the months or years to come. Meanwhile, as long as you address the standard, you are more or less free to be as creative as you wish. Remember however that the writing of tier two documentation is *optional* and is not required by any of the ISO standards. Having reviewed some examples of tier two documentation, let us look at some examples of tier three documentation.

7 Tier Three Documentation

Third level documentation concerns work (or work related) instructions. The documentation of work instructions which might affect production (or perhaps assembly), operation and installation of equipment, etc., is specified in Clause 4.9, *Process Control*, of ISO 9001/Q91 (particularly 4.9.1 (a)). How much and what to write seem to be the two major roadblocks facing foremen or supervisors in charge of tier three documentation.

Some Hints on Documentation

Once responsibilities have been assigned for each paragraph, the following guidelines can be helpful when starting a tier three documentation process:

- Verify if any other paragraph(s) are cross-referenced
- Read corresponding ISO 9004 paragraph(s) for guidelines.
- Re-phrase each sentence of the ISO paragraph into a series of questions:
- Ask the following questions:

"Do these questions apply to 'our' industry?"
("Is it a 'shall' paragraph?")

If YES	If NO
How do we currently address these questions?	Why not? If not applicable: Brief explanation of why not.
Interview person(s) most closely associated with the issues.	If not feasible: Why not (explain)?

Suppose you have to address paragraph 4.18 of ISO 9001 (Training), how would you proceed?

Paragraph 4.18 is cross-referenced with paragraph 4.16 (Quality Record) and paragraph 4.1.2.2 (Verification, Resources and Personnel).

Paragraph 18.0 of *ISO 9004* offers quite a few suggestions on what needs to be done to satisfy 4.18. Having read all the relevant paragraphs, ask yourself "Is this pertinent to my industry?" The answer should be "yes" since paragraph 4.18 applies to any industry.

Proceed by asking the following set of questions:

> "What procedures are currently in place to identify training needs and provide training for all personnel performing activities affecting quality?"
>
> "Are records kept of training?"
>
> "Who can help us answer the above questions?" (Human resources, supervisors, others?)

[Note: If, as is the case with some industries, no or inadequate procedures or records exist, the next step would be to decide who should devise some short, simple easy to implement procedures.]

How Not to Do It

When documenting procedures, the writer should avoid vague statements such as "*If the temperature is tolerable, scales should be OK.*" This seemingly innocuous statement should generate a few questions from an astute auditor. For example, one would like to know:

- What is meant by "tolerable" temperature? How is it defined? How does an operator know what is tolerable? Would "tolerable" have the same meaning to all operators? Could different interpretations of "tolerable" affect the process? If "yes," how? If "no," why not?

[Note: I was once assured by a supervisor of a glue factory that the fact that a temperature gauge on one of the pots had been broken for several years was not a problem. He could always "tell" the pot's temperature by simply feeling the pot with his hand. If the pot was "too hot" (not "tolerable") he would burn himself and the pot had to be cooled quickly to prevent the glue form boiling over! A costly procedure. Is anything wrong with that procedure?]

- The second half of the statement ("*scale should be OK*"), is an excellent example of how NOT to write procedures. The reader is not even assured that the scale IS OK but rather that it SHOULD be OK!

Statements written in the following styles should also be avoided.

Failure of the ethylene unloading pump seal has occurred in the past and is definitely a hazardous problem. The attending unloader should be aware at all times during unloading that any leak can occur. Any presence of vapor, in excess, around the unloading pump should alarm the unloader to shut procedures down until the problem is identified and corrected.

What is wrong with that procedure? Is this a procedure?

Finally, try to avoid writing a manual or procedures in the style quoted below:

The Quality System requires that all functions which affect product quality and consistency, or conformance to customer's requirements be performed by personnel who are adequately trained to perform their respective jobs. This training must be provided by individuals...(etc.)

Reading the above paragraph one has the feeling that the writer did not really know what to say and merely rephrased some of the ISO paragraph. For example, instead of writing "*This training must be provided by individuals...,*" (an obvious paraphrasing of ISO), why not simply write that "*Training is provided by qualified individuals.*" Naturally, one would have to define what is meant by qualified individuals (e.g. months or years of experience, certification, special training, etc.). Records of all training must also be kept.

Some Examples of Effective, Easy to Write Third Tier Documentation

Contrary to what is generally believed, the documentation of procedures need not always be an elaborate enterprise. The following three examples demonstrate how simple procedures can be concisely written. Each example, modified from actual procedures, took less than an hour to write. Naturally, more complex procedures would take longer than an hour to document. However, if a procedure becomes much too complex and requires input from several individuals, perhaps it should be broken down into sub-processes. It is unfortunately impossible to define the "right" size for any particular process. This must be "felt" by each process team.

Example 1: Checklist for Unloading Tank Cars

<table>
<tr><td colspan="2">Procedure: Ethylene Tank Cars</td><td colspan="2">Last Update: January 19, 1990</td></tr>
<tr><td colspan="2">Written By: Joe Geisner</td><td colspan="2">Approved By: Bill McIntire</td></tr>
<tr><td colspan="2">Signature:</td><td colspan="2">Signature:</td></tr>
<tr><td>Date:</td><td>Car #:</td><td>Unloaded by:</td><td>Time in:</td></tr>
</table>

Initials

______ 1. Test car vacuum, liquid level and pressure.

______ 2. Attach ground cable to tank car.

______ 3. Ensure that nitrogen flows through pump seal.

.

.

.

______ 40. Disconnect ground cable.

______ 41. Check defect card holder for defect card.

Example 2: Unloading procedure for trucks

(1) Park trailer sloping to back. Measure slope at flange on hatch with level ______.

(2) Have driver drive truck to plant entrance and wait for us to unload ______.

(3) Bring tools _____, gaskets _____, and protective clothing.

This last example is a clever way of combining the writing of procedures with training. One additional benefit of such an approach is that it allows the "instructor" to verify and update the effectiveness of each procedure.

Example 3: Combining Training with Procedure Writing

Procedure: Waxing	Last Update: January 19, 1988
Written By: Jim Allison	Approved By: Richard Thomas
Signature:	Signature:

	Operator's Initials	Assistant Initials
1. Pump for oxidizer #2	______	______
2. Check temp. for wax (state temperature)	______	______
3. Start-up of wax circulation	______	______
4. Etc.		

A copy of the above procedure could be sent to Human Resources to document training on the "waxing" procedure.

How to Document Complex Procedures

If there is more than one way to operate a process, say so in your procedures. The actions taken on certain procedures often depend on a variety of conditions. For some processes, documenting each and every possible option could require as many as two dozen or more procedures, some of which would occur very rarely. In such cases the Pareto principle often takes effect, that is, a few procedures usually cover 85 percent of all possible actions. Define these standard procedures which account for 85 percent of your actions (corrective or otherwise). Explain that in some cases (fifteen or twenty percent of the time) deviation from these procedures will occur. However don't simply state that deviations do occasionally occur, explain in a few brief statements what is/are the deviation procedure(s). You could for example explain that for each deviation the operator or person in charge must:

- Inform supervisor.
- Inform Action Team.
- Brainstorm decision.
- Document and file corrective action taken or suggestion(s).
- Implement corrective action(s).

A similar approach could be adopted by the maintenance crew, who as we all know, seems to always be stretched to the limit. Maintenance crews often claim that they do not have time to maintain detailed records. However, detailed records are not necessarily required. If a check list is prepared and maintained, documentation efforts should not require more than a couple of minutes. Such records can be invaluable for preventive maintenance. If designed properly, they can facilitate the task of estimating Mean Time Between Failure (MTBF) or detecting patterns of break-down.

An aluminum plant in the Pacific Northwest has adopted a very impressive set of maintenance procedures. Upon visiting the maintenance facility you immediately sense that something is "odd" about the place. The first thing you notice is that the facility is extremely clean and newly painted. When you next notice the green plants and the art posters, you know you have just entered a unique environment. Maintenance records for each piece of equipment have been computerized. These records include:

- When the equipment was inspected.
- What was inspected (checklist is included) and by whom.
- What measurements were made and what were the readings.
- What components/parts were changed/replaced and why.
- Next inspection due date.

Data can be entered to a mainframe from any one of several remote P.C. located throughout the plant. A very impressive working system indeed!

More Examples

Note: *Comments in italics or question marks are notes written by the consultant in charge of preparing the work instruction. All comments will have to be addressed. SOP stands for Standard Operating Procedure. Only the first few steps have been included.*

Tank Car Wash Rack and Inspection Procedures

1.0 Introduction

Prior to loading, most tank cars are washed, and internally and externally inspected. Cars that are recycled for specified products (*list*) are not washed. These recycled tanks are controlled (*how*) by the shipping services supervisor.

Procedures:

2.0 Preparation

1. Switching crew selects tank car.
2. Secure tank car by chock *and* hand brakes.
3. Connect ground cables to each car.
4. Attach safety appliances to each car.
5. Depressurize all pressurized tanks.
6. Manway bottom outlet valves.
7. Open: Secondary valves.
Vent valves.
Induction valves.
8. Remove cap from bottom outlet valve nozzle.

Document no.: TKWR02	Description: Tank car wash and inspection procedure
Revision no. : 0	Sheet: 1 of 3
Prepared by: Paul Brown	Approved by: T. Jones (signature)
Issued by: Cleaning Section	Issue Date: May 31 1991

3.0 Washing

9. Wash car with hot water.
10. Purge car with hot air blower.
11. Service supervisor must then check for:

Oxygen content (*how*?).
Temperature (*define*?).
Smell (*for what?*).
Combustible (*how*?).

12. If any nonconformities are found service supervisor issues correction notices (*verbal or written*?)

4.0 Internal Inspection

(*Internal inspection consists of five steps (13-17) which would have to be explained. Step 15 includes such words as "excess water," " excess rust," etc. Try to avoid the use of such vague expressions. What is considered "excess water or rust"? How is it defined or quantified?*

13. (Explain as required)

14. (Explain)

15. The following is checked by visual inspection:

16. (Explain)

17. (Explain)

Document no.: TKWR02	Description: Tank car wash and inspection procedure
Revision no. : 0	Sheet: 2 of 3
Prepared by: Paul Brown	Approved by: T. Jones
Issued by: Cleaning Section	Issue Date: May 31 1991

5.0 External Inspection

(Internal inspection consists of eleven steps (18-28). The person(s) closest to the procedures should have a say as to how the "job gets done." Try to avoid having managers or foremen write all procedures without the assistance of operators. Be as concise as possible. Remember! Only write down what you actually do, not what you think is done. If you would like to see a procedure modified, you must first make sure that the modified procedure is accepted, understood and applied. Do not document fictitious procedures.)

18.

19.

20.

21.

22.

23.

24. Correct any leak found.

25.

26.

27.

28. Remove car from wash rack.

Next step: The manager must review the above description for inaccuracies and address all comments.

Document no.: TKWR02	Description: Tank car wash and inspection procedure
Revision no. : 0	Sheet: 3 of 3
Prepared by: Paul Brown	Approved by: T. Jones
Issued by: Cleaning Section	Issue Date: May 31 1991

Tank Car Wash Rack and Inspection Procedures
Version 2

Document no.: TKWR02	Description: Tank car wash and inspection procedure
Revision no. : 0	Sheet: 1 of 3
Prepared by: Paul Brown	Approved by: T. Jones (signature)
Issued by: Cleaning Section	Issue Date: May 31 1991

Note: Rather than include the document control information at the bottom of each page, some documents have it at the top of each page. Luckily, thanks to sophisticated word processors, any format can easily be adopted. The above box was prepared using the "Insert Table" *option in the MacIntosh Word software. Only the first of three pages is reproduced here.*

1.0 Introduction

Prior to loading, most tank cars are washed, and internally and externally inspected. Cars that are recycled for specified products (*list*) are not washed. These recycled tanks are controlled (*how*) by the shipping services supervisor.

2.0 Preparation

1. Switching crew selects tank car.
2. Secure tank car by chock *and* hand brakes.
3. Connect ground cables to each car.
4. Attach safety appliances to each car.
5. Depressurize all pressurized tanks.
6. Manway bottom outlet valves.
7. Open: Secondary valves.
   Vent valves.
   Induction valves.
8. Remove cap from bottom outlet valve nozzle.

Etc.

Conclusions

As with second tier documentation, there are no third tier writing style or format standards to follow. You can be as creative and as detailed as you wish. As long as you are confident that what is written down is indeed what is done, you should encounter no difficulties. Remember that you need not write procedures for everything you do (re-read clause 4.9 of Q91). In some of the plants I have visited, detailed work instructions were automated. Other plants have implemented very effective computer control procedures which allow the process engineer to monitor a process for as many as a dozen or more parameters. Monitoring includes on-line statistical process control charts of a selected set of key variables. All of these procedures are acceptable as long as they are used effectively by operators or process engineers to monitor product quality. This may seem a trite statement but the author has seen too many "sophisticated" programs — written by overzealous but well-intentioned programmers from the Management Information Systems department — designed to gather gigabytes of data for no apparent purpose.

8 Documenting Your Procedures

One popular but not necessarily universal method to document processes is to flow chart them. One of the advantages of flow charting is that it forces a group of people (hopefully, the process owners) to describe or brainstorm their conception of the process. Once a consensus has been reached on the process' flow, opportunities to improve can and should be suggested. Documentation must not become a static event rather, it should be dynamic and allow for continuous improvement in process effectiveness.

How to Document Procedures

Follow Deming's **P**lan, **D**o, **C**heck and **A**ct cycle. Above all:

1. *Say what you do.*
2. *Do what you say (procedures).*
3. *Record that you have done it.*
4. *Audit for compliance and effectiveness (internal audits).*
5. *Feedback and continuously improve.*

Preliminary steps:

a. *Define the boundaries of the process/procedures.*
b. *Determine whether all persons responsible for the process/procedure are present.*

Once these steps have been taken you can proceed either by:

a. Interviewing all responsible persons, write down what they say and have them review what you wrote until you have written down what they really do or,

b. Assemble all responsible persons and proceed as outlined below.

1. Brainstorm activities (5-10 minutes).
2. Review ideas submitted during brainstorming session.
3. Prepare a new list of activities (consensus).
4. Attach a flow chart symbol to each activity.
5. Flow chart process.
6. Review, and make "final" improvements.

Note: During all stages remember to ask:

What is the purpose of the process/procedure?

Who is/are responsible for the process?

Who is/are the next customer(s) downstream and is/are the last supplier(s) upstream?

When does the procedure apply (always, sometimes, exceptions)?

How the procedure operates (i.e., technical issues). These should be included in the work instructions procedure manuals.

Is it a key procedure that affects the product's quality (see Chapter 10 question 4).

How to Flow Chart

One of the easiest ways to document a process or a procedure is to flow chart it. A flow chart is a schematic representation of a process/procedure. There are basically six symbols used to depict various actions (see Figures 8.1, 8.2 and Table 8.1). Since there are unfortunately no standards regarding the use of flow chart symbols, confusion can arise especially if you are not careful defining your symbols. The following "definitions" reflect only the author's preference.

The square usually means an operation or task. The diamond means that a decision point has been reached. The branching out of a diamond is invariably a "yes" or a "no." The circle can mean inspection. The inverted triangle means storage of anything (e.g. data, liquid, parts, information, etc.). The arrow indicates flow or transportation of material or information or anything else and the rounded rectangle depicts some delay in the process.

The brainstorming session and the resulting flow chart for the process "*Preparing a Cup of Instant Coffee*" (Table 8.1) are found on the following pages. The easiest procedure to follow is to first brainstorm all the activities, rank them in a logical sequence of events and attach the appropriate flow chart symbol to each activity. The last step would consist of connecting all the symbols with arrows, thus indicating the process flow.

Figure 8.1 Some Typical Flow Chart Symbols

(Some authors assign different definitions to the symbols)

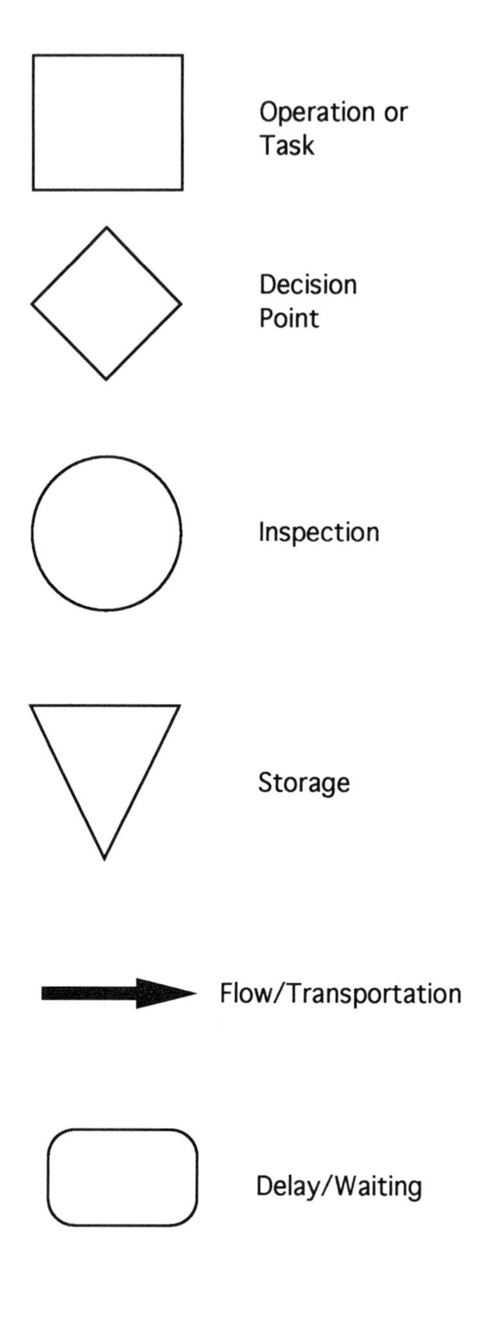

Table 8.1 Process: Preparing a Cup of Instant Coffee

Brainstorm session	Symbol
1. Get instant coffee	
2. Get coffee pot	
3. Add water to coffee pot	
4. Add desired amount of instant coffee to cup	
5. Heat water	
6. Wait for water to boil	
7. Hot enough?	
8. Pour water in cup	
9. Turn off plate	
10. Drink coffee	

How could you improve the above process?
Any missing steps?

Note: How would you improve the "*Preparing a Cup of Instant Coffee*" to anticipate the possible addition of sugar and milk? (Hint: You will need two additional decision points).

Figure 8.2 Preparing a Cup of Instant Coffee

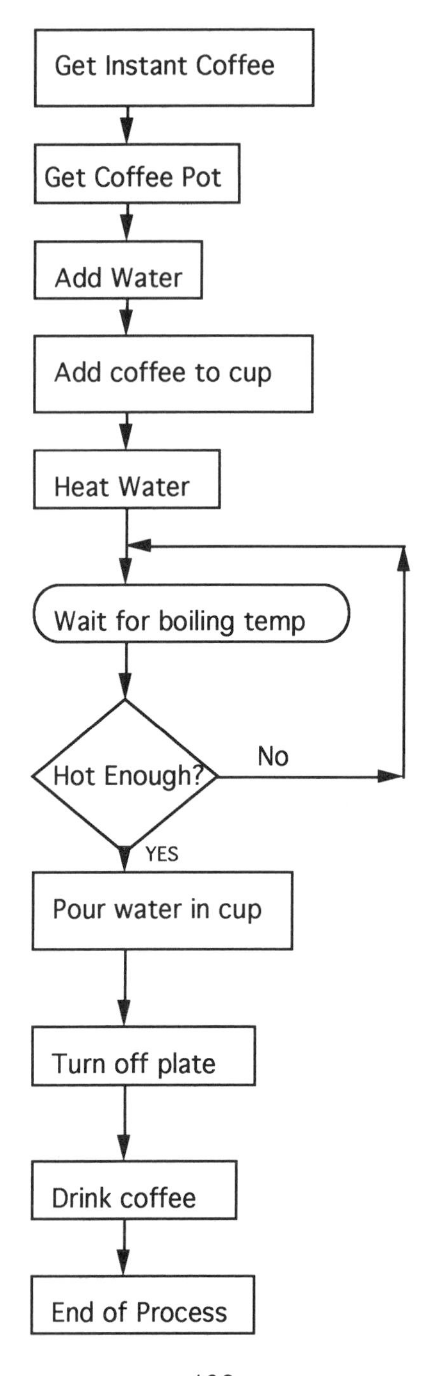

Flow Charting Exercise 1

Note: The following is a verbatim transcription of an incoming order procedure. Your task is to: (1) Prepare a flow chart for the incoming order procedure, (2) identify which ISO 9001 or 9002 paragraphs (if any) are covered by this procedure and (3) comment on the process from an auditor's point of view (i.e., what questions or clarifications would you ask?).

The Shipping Clerk receives the incoming order on an IBM Datacom printer centrally located in the Traffic Office. Once the order is received, it is reviewed to determine the following:

(1) Product to be shipped.
(2) Amount of product requested.
(3) Mode of transportation to be used.
(4) Date of shipment and requested delivery date.
(5) Determine if specialized equipment is needed.
(6) Has the customer referenced any specific shipping instructions?

The Distribution Specialist determines product availability, sets priorities, and orders specialized equipment.

The order is filed by product name in an "Open Order" book. The order is later pulled according to its shipment date. A daily "Loading Instruction Sheet" is prepared by the shipping Supervisor comprised from information entailed on the Open Order Registrar, internal supply and demand, etc.

Open orders are pulled on a daily basis and the information is coordinated with the "Loading Instruction Sheet". A worksheet is prepared by the Distribution Clerk which included billing information pertaining to each individual order that will be shipped for that day. The worksheet also reflects the loading information such as container number, seal numbers, weights, etc., received from the Shipping Service area.

Once the vessel (tank car/tank truck/barge) has been properly loaded and the loading information given to the Distribution Clerk, the tank cars, for example, may then be weighed by the Contract Switching Crew. The weights are received and printed in the Traffic Office. The weigh ticket will distinguish the container number, as well as the gross, tare, and net weights of that container.

The weights received on the weigh ticket are entered into an automated billing system to acquire a Bill of Lading for movement of that shipment. Some of the information reflected on the Bill of Lading is as follows:

(1) Date of shipment and B/L number.
(2) Customer name and destination.
(3) Product and description.
(4) Carrier name and routing.
(5) Weights and freight amounts.
(6) Internal and customer order numbers.
(7) Seal numbers.
(8) Special instructions.

After the Bills of Lading have been printed for each shipment, the automated billing system will prompt the systems to print "Empty Return" Bills of Lading for each shipment. These "Empty Returns" will be used by the customer or returned (empty car) to the proper shipping location.

The distribution Clerk will then distribute the various copies of the Bills of Lading as follows:

(1) One copy will be faxed to the designated railroad carrier.
(2) One copy will be sent to the Switching Crew to begin movement of the tank car from the Plant.
(3) One copy will be attached to the Empty Return and mailed to the customer along with a copy of the weight ticket.

The Distribution Clerk also maintains an Inventory Sheet which is updated, as needed, to reflect any loaded tank cars currently being held in the Plant for future shipment. At the end of the day, the Distribution Clerk reports into the system all shipments that were shipped for that day to acquire a "Shipped Report." The shipped Report and the Inventory Sheet are then distributed to various Control Rooms throughout the Plant to be used accordingly.

Flow Charting Exercise 2

Note: The following is a verbatim transcription of a loading (rail car) procedure. Your task is to: (1) Prepare a flow chart for the incoming order procedure, (2) identify which ISO 9001 or 9002 paragraphs are covered (if any) by this procedure and (3) comment on the process from an auditor's point of view (i.e., what questions/clarifications would you ask?).

The loading section receives a loading instruction report along with verbal information each morning. After the loading instructions are verified, the loaders will do a pre-loading inspection prior to the start of loading. The Rackmaster will calculate the amount of liquid to be loaded into each car. He also records the due dates on the Consolidated Loading Report, tank car, safety valves, heating coils, DOT classifications and gaskets on the cars. The tank form loader will gauge record temperatures and line-up each tank to the proper loading rack. After purging with N_2, each car will be checked for Oxygen content. When rail cars have been made ready, the loading arms are installed and hoses are attached to the proper vent headers. Each car is filled to the predetermined level, sampled and disconnected from the rack. Then the cars are checked for leaks and secured. Seals with attached product labels are installed on the top and bottom of each car. Placards are placed in the holders on all four sides of the cars.

The samples are sent to the plant laboratory for product quality certification. Re-samples are needed if the analytical results show the product to be off-spec. Rail cars that are off-spec will be rejected and off-loaded at a later time. Rail cars are weighed in at the plant and the overweight cars are returned to the rack to remove the excess gallons.

All approved cars are shipped with the Bill of Lading. All tanks are blocked-in, gauged and temperature recorded. The consolidated transfer report is made out and checked for a material balance. Original copy is sent to the owning unit and one is kept for the shipping files.

9 How to Proceed?

The following recommendations are but a brief listing of questions that should be asked *and hopefully answered* by members of the ISO registration team prior to assigning tasks. It has been the author's experience that, in too many cases, companies tend to aimlessly rush towards registration only to later have their momentum and enthusiasm diminished by countless nagging difficulties. Failure to plan, prepare and anticipate problems can lead to unnecessary and frustrating delays. The following is a list of questions and suggested strategies that could be discussed by the ISO committee over a period of two to three meetings.

What to discuss?

- Who will be responsible for monitoring the implementation efforts? How will employees be notified of the company's desire to achieve ISO certification?

- Which plant, plants, division or product lines will be registered?

What to discuss?

- Who will do what?

How to achieve

One of the easiest ways to assign responsibilities is to prepare a responsibility matrix. The matrix (see next page) helps the quality manager/director or the committee in charge of implementation, decide who shall/should be responsible for which ISO paragraph(s).

Responsibility Matrix

Functions or Departments	Plant Manager	Quality Manager	Operation	Purchasing	Lab. Mngr.	Marketing	R&D	Others
4.1.1 Quality Policy								
4.1.2 Organization								
4.1.2.2 Verification Resources								
4.1.2.3 Management Representative								
4.1.3 Management Review								
4.2 Quality System								
4.3 Contract Review								
4.4 Design Control								
(All sub-paragraphs)								
4.5 Document Control								
4.6 Purchasing								
4.6.2 Assessment								
4.6.3 Purchasing data								
4.6.4 Verification of Purchased Prod.								
4.7 Purchaser Supplied								
4.8 Prod. Identification								
4.9 Process Control								
4.10 Inspection+Testing								
4.11 Insp., Measuring, Test								
4.12 Inspection+Test Status								
4.13 Control of Nonconforming								
4.14 Corrective Action								

What to discuss?

- Who will conduct the pre-assessement audit (in-house team, consultant or combination)? When should the pre-assessement audit be conducted?

How to achieve

Pre-assessement audit findings can help better determine how long the implementation process will take. Refer to Appendix C for procedures.

What to discuss?

- Decide on who will write the quality manual and who will need to *review* it. Will the manual be written by one individual or will sections be "assigned" to various persons?

How to achieve

There are essentially two options: either the quality manager writes the quality manuals or he/she writes a few paragraphs and delegates sections to volunteers or "assigned volunteers." The first option tends to be more expedient simply because one person is in charge of the document. When sections of the manual are assigned to various writers, delays tend to be in direct correlation to the number of writers! On the positive side, one should note that the writing process is not only more democratic but probably more accurate.

In many cases, the responsibility to complete the quality manual usually falls upon a quality manager/director. This responsibility needs to be shared with the highest ranking executive. Such sharing of responsibilities indicates a total commitment to quality. If direct responsibility cannot be obtained, the highest ranking executive should at least review *and approve* the final version of the manual. This can be achieved by having each page of the manual - written by the quality manager/director - signed by the president of the company or plant manager for example (see Appendix A). Other options would include having each person directly responsible for a particular ISO clause or set of clauses approve the relevant section(s).

Since the quality manual *is* a summary of a company's quality system, it must be reviewed, prior to release, by each division directly affected by the manual's contents. To achieve that goal, a distribution list must be prepared to include all directors/managers involved with the quality manual's contents. That distribution list will become part of the manual. In most cases, the list will not exceed 15 names.

Circulation of the quality manual is essential since it allows everyone to participate, communicate or otherwise share in the development and formalization of the company's quality system. Moreover it allows for the coordination of efforts between each division, particularly the coordination of tier one and tier two documentation. Once the final version has been edited, the manual should be submitted to the president for final editing and approval.

What to discuss?

- Decide who will write tier two and tier three documentation.

How to achieve

There are several strategies available. Directors or their delegates could organize the tier two effort. Using the internal supplier/customer model, a division could document its tier two and have its internal supplier(s) verify/audit the procedures.

Tier three documents could be delegated to supervisors or even operators. One useful method to adopt is to have the procedure (tier two or tier three) written by one individual and verified by the appropriate manager. Remember to write what you are *currently* doing not what you should be doing or *will* be doing in the future. I do not mean to suggest that continuous improvement is not part of the ISO registration process, on the contrary. I merely want to warn people who want to use the registration process as a means to "catch-up" on everything that needs to be done or should have been done years ago. Documenting what you are currently doing will be demanding enough. Accomplish that task first and then move on to continuously improve your system. Attempting to do everything at once might well enhance your chances for *failure.*

What to discuss?

- How will documents be controlled? Who will manage document control (one or more person(s))? Which documents will need to be controlled (quality manual, process documentation, quality records, etc.)? Remember the importance of paragraph 4.16 of ISO 9001 (4.15 for ISO 9002)! Also remember to keep procedures simple and concise.

How to achieve

See examples in Chapter 5. If your plant has a laboratory, you might want to inquire as to how the laboratory controls its procedures.

What to discuss?

- How and who will conduct internal audits?

How to achieve

A *brief* procedure will have to be written (frequency, documentation, how are nonconformities resolved?). What training requirements will need to be addressed? (Remember paragraph 4.1.2.2 of ISO 9001/Q91.)

What to discuss?

- Prepare a preliminary timetable and update as often as necessary (use input from pre-assessment audit). Document how long the implementation record does take. This will allow you to document part of your internal cost.

What to discuss?

- Finally, how will you attend to the training needs as specified in paragraph 4.18 of ISO 9001?

Naturally, you will not be able to address all of the above questions in one meeting. It might well take two to three meetings to sort out.

How Much Time?

When asked "how much time will it take to implement a quality assurance system which will conform to the ISO model?", the answer can only be, "it depends." Indeed, it depends on the size of the company, the willingness or motivation to achieve registration by a certain date, the consultant's expertise in guiding the client and avoiding mistakes, the amount of documentation required, who will write the documentation, how much resistance will there be from the various organizations, how much delegationwill there be, how many people will (willingly) participate, and other intangibles.

It is estimated that for a small company (i.e. up to 200-300 employees) having only a rudimentary system in place, 100-140 days (not necessarily consecutive) would be required to bring the company up to ISO standards. Most companies achieve the task within nine to fourteen months. The basic rule would be to allow: twelve to fifteen days for the *actual writing* of the quality manual (many more weeks might be needed to decide on what will be included in the manual), at least half a day per tier two documentation (allow for approximately thirty to forty five days), half a day per work instructions and the balance (usually anywhere from forty to fifty-five days, depending on plant size and number of nonconformities found) for auditing and testing the system (see Table 9.1).

These estimates assume that the company in question is totally committed to the task at hand and is willing to invest the necessary resource to achieve ISO registration within a reasonable time frame (i.e. nine to fourteen months). Such is not always the case. I have seen companies caught in the midst of reorganization spend as much as one year on the quality manual alone! One should also note that the above estimates *do not include any training.* This is not to imply that training will lengthen the registration process but rather, that it will increase the cost.

Training can easily consume as much as seventy percent of an allocated budget. Since technical and quality related training sessions are required by ISO, OSHA, EPA and other agencies, it is difficult to avoid. The difficulty with training programs is that they need to be carefully timed in order to be effective. They also need to be practical rather than theoretical.

Implementation Schedule

The implementation schedule listed in Table 9.2 gives an overview on how a small to medium organization (100-500 employees) can plan a nine to eleven month program. The schedule assumes a stable company, that is, a company which is not experiencing multiple reorganizations. Companies who are in the midst of a major or even a minor restructuring can anticipate (in a worse case scenario) spending as much as twelve months on documenting the quality manual alone!

The *Quality Awareness Campaign* is an essential, and yet often ignored, first step. The purpose of the awareness campaign is to inform the workforce of the company's desire, reasons and objectives to achieve ISO 9001, 2 or 3 registration within a certain time frame. Although many companies would agree that an awareness campaign is a good idea, few actually bother to practice what they preach. Some of the more committed companies would devote a one or two day in-house "ISO 9000 Awareness" seminar designed for their upper management. Others send one or two representatives to one of the many ISO 9000 public seminars. Having attended the seminar, the delegate(s) is/are then expected to embark on the "ISO registration program." Upon returning to their respective plants, these delegates will find out that the few who know what ISO 9000 is all about will probably think that ISO registration is a good idea *if only they had the time to help*! As for the others, who don't even know about ISO 9000, much convincing will have to be done to gain their support. Such frustrating experiences can be mitigated if upper management were to devote some efforts towards a well planned, informative campaign.

Table 9.1 Estimated Time For ISO 9001/9002 Implementation				
Tier	**Originator**	**Activity**	**Time**	**Months**
Quality Manual (1st tier)	Quality Manager or team effort QM will have to interview individuals to ensure accuracy.	Organizing, Interviewing etc. Writing, Reviewing, correcting (two or more cycles) Final print Total =	5 days 5 days 3x3 = 6 days 1 day 15-17 days	Activities spread over two to three months.
Second tier: Who does what with respect to each of the ISO paragraphs.	Quality manager OR Department head OR joint effort.	Use format of •Purpose •Scope •(References) •(Definitions) •Responsibilities •Procedures	Half a day per section. Total should not exceed 30-45 days.	Activities may be spread over twelve to sixteen weeks.
Third tier: Work Instructions and Process Flow charts. Emphasizes how job is done. Can be used for training.	Supervisor with assistance of workers or employees.	Can use flow charts or written instructions of how job is performed.	1 day to write plus 1 day to edit. Allow for 40-55 days. ------------------ Total = 85-105 days of documentation plus 15-20 days of Internal Audits and corrective actions for a Grand Total of 100-137 days.	Activity may be spread over three to four months. ------------------ See schedule (Table 9.2) for activities.

ISO 9000 Implementation Plan (<300 Employees)

ISO Awareness Campaign
- Establish Timetable
- Communicate to All the ISO 9000 Implementation Plan

Nominate ISO Representative(s)/Coordinator(s)

Form Implementation Teams
- Quality Manual (Who will write what? Who will edit? GANTT)
- Process/Procedures Teams

Begin Writing Quality Manual
- Review ISO 9004 Guidelines
  - How to organize, how to update, how to write, etc.?
- Coordinate with Process/Procedure Team
- Determine Which Procedures Will Have to be Referenced

Develop Process Flow chart (optional)
- Define Boundaries of Processes
- Identify Which Process(es) Will Have to be Flow charted
- Identify Procedures that Will Have to be Documented

Document Tier Two and Three Procedures
- Set Up Interviews
- First Draft of Procedures
- Verify the Accuracy of Written Procedures
- Identify Special Work Instructions in Need of Documentation
- Laboratory Procedures!

Quality/ISO Audit Training
- Problem Solving Techniques
- Statistical Techniques

Set Up Pre-Registration Audit

Monitor Implementation Process (On-going)
- Begin Internal Audits, Document Corrective Actions

Schedule Compliance Audit

Clear Discrepancies

Registration

Following the awareness campaign, "ISO 9000" representatives will have to be appointed. The necessity of this action does not require any further explanation except to state that without a steering action group (I hesitate to call it a committee), whose purpose will be to organize, monitor, catalyze and energize the implementation process, the likelihood of success is greatly diminished.

Table 9.2 Estimated ISO 9000 Implementation Schedule For Less Than 500 Employees

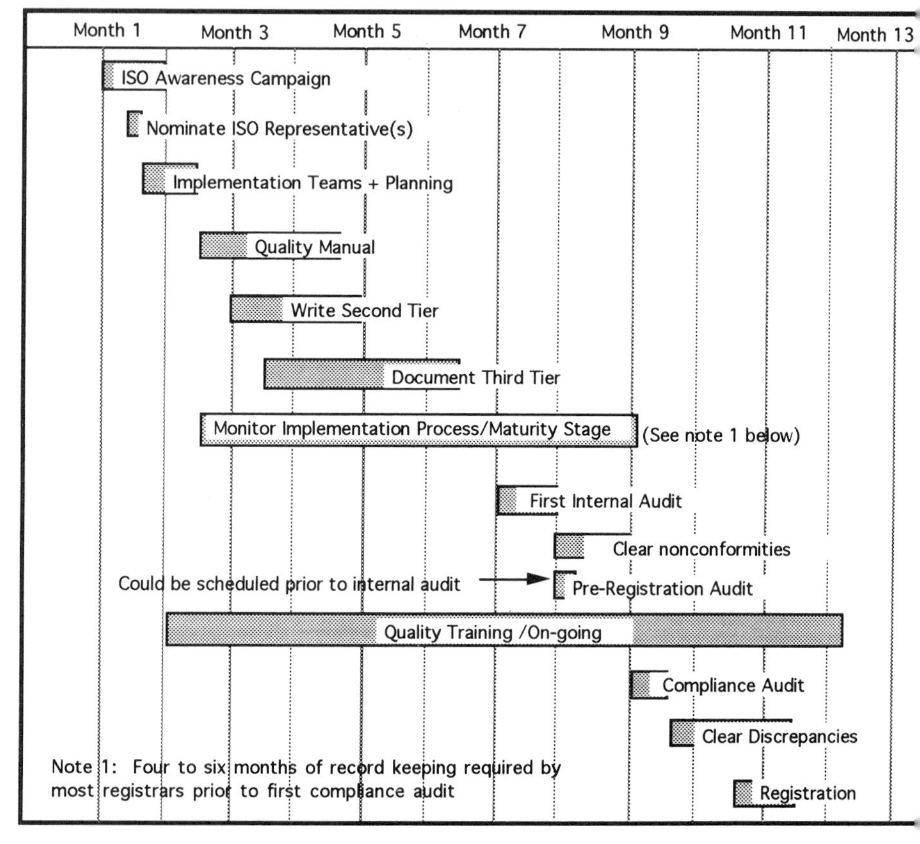

Once the documentation assignments have been allocated, the system will have to be constantly monitored/assessed to ensure effective and steady progress. As the system matures, the time will come (four to six months down the road), to conduct the first internal audits (see Chapter 13 for advice on how to conduct internal audits). This is a crucial step in the implementation process. Indeed, in order to properly test the system, a certain amount of time will have to elapse. Most registrars will not even bother to assess/audit your company until your quality assurance system has been in place for approximately six months. Naturally, you need not wait six months to conduct your first audit.

When all seems in order, third party auditors can be invited to officially assess your readiness. Assuming all goes well, you should receive your certificate within forty-five to sixty days (for further details on third party audits, see Chapter 12).

The Cost of Certification

The estimates listed on the next page (Tables 9.3 and 9.4) are based on 1991 costs. They are only provided as guidelines to help the reader estimate his/her implementation costs. Without first conducting a preliminary audit, it is extremely difficult, if not impossible, to intelligently estimate how much a company will have to invest to be ISO 9001, 9002 or 9003 registration ready. Some companies have invested as much as a $100,000, others much less and still others substantially more. There are much too many parameters to consider when estimating costs, including cost measurement and definition itself. Excluding training and preparation (i.e. set of meetings leading to actual implementation for example), the cost figures presented below are calculated for a small to medium size plant (100-500 employes). As the reader can verify, the registration and implementation costs for a 100-300 employee site can range anywhere from $26,000 to $45,000 (1991).

Table 9.3 Estimated ISO 9000 Certification Costs (1991)
(Author's estimates based on one (100-300 employees) site visit only.)

	Man-days	Rate	Cost
Certification Costs			
One-time			
Pre-registration Audit Fee *	1	$1,200/day	$1,200
Certification Audit Fee*	4-12	$1,200/day	$5-15,000
Total One-time Annual			$6-16,000
Registration Fee. Applies only to some registrars.	N/A	Variable	
Bi-Annual Surveillance Audit**	2	$1,200/day	2,400
Total Annual			$2,400

* Audit fees do not include expenses. Pre-registration fee based on 1 auditor for 1 day. Rates are per auditor.
** Surveillance audits vary from registrar to registrar. Some registrars conduct a complete assessment every three years. Others conduct partial assessment once or twice a year.

Table 9.4 Estimated ISO 9000 Implementation Costs (1991)
For a Small (100-300 EMPLOYEES) SITE
Exclusive of Preparation Costs (meetings, organization, etc.)

	Man-days	Rate	Cost
Implementation Costs			
One-time			
ISO 9000 Quality Manual *	15-20	$25/hr	$3-4,000
Tier Two*	30-45	$25/hr	$6-9,000
Tier Three*	40-55	$25/hr	$8-11,000
Internal Audit +Discrepancies*	15-20	$25/hr	3-4,000
Total One-time	100-140		$20-28,000
Quality Training **	300	$25/hr	$60,000

* Salary/benefits $52000/year.
** 20% work force trained 3 days/year. Salary/benefits $52000/year.

Conclusions

The cost and time estimates presented in this chapter are based on the assumption that a company is dedicated to the ISO 9000 registration process. Dedication means, among other things, that proper resources, training and planning must be allocated to facilitate, encourage and ensure success. In many cases, the ISO 9000 registration effort becomes one more "program" that must somehow be squeezed in between current activities. Certainly, no organization can neglect the very activities that keep it in business while focusing on the ISO standards. However, failure to properly organize for ISO registration will certainly lead to unnecessary delays and frustrations.

There has been a considerable amount of misinformation relating to the cost of obtaining ISO 9001, 2 or 3 registration. The author of an article published in *Industry Week* (August 19, 1991, p. 57), claimed that a facility of 250 people would have to spend about $500,000 to obtain registration. Assuming that the quoted figure is not a typographical error, $500,000 would buy approximately 500 auditors/day! Enough to audit the likes of Ford, GM or Boeing but at least twenty-five times more than what would be required to audit a 250 man operation. Such outrageous claims are most unfortunate for they are quickly picked up by opponents of ISO 9000. Indeed, the $500,000 audit story was mentioned in a November, 1991 newsletter shown to the author during one seminar. The twelve page newsletter, replete with inaccuracies about the ISO 9000 series, went on to conclude that ISO 9000 certification and regulation (*sic*) "is superfluous and unjustified."

Having reviewed some of the more important organizational and cost related issues, we shall now turn our attention to some practical and often asked questions regarding the ISO 9000 series.

10 Often Asked Questions

Despite my continuous efforts to explain the scope and field of application of the ISO 9000 series, a particular set of recurring questions seem to be asked. They are:

(1) "Must we inspect all raw materials coming into the plant?"

(2) "How much detail do I need to include in my process procedures? Must I explain everything?"

(3) "How many procedures do I need to write?"

(4) "Since everything affects quality, must I document everything?"

(5) (Statement) "I am currently writing all of my maintenance procedures (several hundreds of them), but I can tell you right now that my guys won't read them!"

(6) "We co-own our processes, how do you handle that?"

(7) "I have over 5,000 instruments, gauges and other pieces of equipment. Must I calibrate and do precision studies on all of them? How many records do I need to keep?"

(8) "My operators know what they have to do because they have done it for fifteen or more years. Do I really need to write procedures?"

(9) "How do you address corrective actions when they depend on the type of problem?"

(10) "Purchasing is not part of our plant. What do I need to do?"

(11) "Some of my customers have asked me to document my costs of quality. Would I have to do the same for ISO certification?"

(12) "How do I determine what parts of my processes need third tier documentation?"

(13) "Will ISO certification help reduce customer audits?"

Let us go over each question.

(1) "Must we inspect all raw materials coming into the plant?"

The short answer is no, but let us refer back to the standards and the Guidelines. ISO 9001 (paragraphs 4.10.1.1 and 4.10.1.2) and ISO 9002 (paragraphs 4.9.1 and 4.9.2) are identical.

> 4.10.1.1 *The supplier shall ensure that incoming product is not used (except in the circumstances described in* 4.10.2*) until it has been inspected or otherwise verified as conforming to specified requirements.* ***Verification shall be in accordance with the quality plan or documented procedures.***
>
> 4.10.1.2 *Where incoming product is released for urgent production purposes, it shall be positively identified and recorded (see 4.16) in order to permit immediate recall and replacement in the event of nonconformance to specified requirements.*
>
> *Note: In determining the amount and nature of receiving inspection, consideration should be given to the control*

exercised at source and documented evidence of quality conformance provided.

One of the key sentences in the above two paragraphs is that "Verification shall be in accordance with the quality plan or documented procedures." In other words ***your*** Inspection and Testing procedures should spell out what product(s) are to be tested according to what test procedure, what products only require a certificate of conformance and what products require no such certificate. The issue is not "what must I test for in order to satisfy ISO," but rather "do we currently have procedures to test incoming materials and how do we currently address the requirements specified in the above two paragraphs?"

The ISO 9004 does provide valuable guidelines on how to interpret the Standards. For example with respect to Product Verification, paragraph 12.1, Incoming Materials and Parts, states:

> *The method used to ensure quality of purchased materials, component parts and assemblies that are received into the production facility* ***will depend on the importance of the item to quality, the state of control and information available from the supplier and impact on costs*** *(see 9.7 and 9.8)* [ANSI/ASQC Q94-1987, paragraph 12.1].

Who decides on "the importance of the item to quality"? You and possibly your customers (depending on your contractual agreement), decide what is important. In most plants I have visited, particularly chemical plants, incoming materials are classified into two or three broad categories. Some, for example, catalysts or other key components are very critical and are either sampled or require a certificate of conformance. Other materials, such as certain chemical products delivered via an extensive pipeline network by suppliers hundreds of miles away, are not even sampled. There is nothing wrong with those procedures as long as they are properly documented, are standard industry practice and are acceptable to you and your clients.

(2) "How much detail do I need to include in my process procedures?"

ISO 9004 addresses several paragraphs to the issue of Process Control and procedures. One of the most pertinent paragraphs is perhaps paragraph 10.1.3 of ANSI/ASQC Q94 which states:

> *Verification at each stage should relate directly to finished product specifications or to an internal requirement, as appropriate.* ***If verification of characteristics of the process itself is not physically or economically practical or feasible, then verification of the product should be utilized.*** *In all cases, relationships between in-process controls, their specifications, and final product specifications should be developed, communicated to production and inspection personnel, and documented* [ANSI/ASQC Q94, paragraph 10.1.3]. [Note: The reader should refer to all other paragraphs particularly those referring to process capability (10.2).]

The standard — Q91 4.9.1 — states that "The supplier shall identify and plan the production and, where applicable, installation processes which directly affect quality and shall ensure that these processes are carried out under controlled conditions." Moreover, "documented work instructions...*where the absence of such instructions would adversely affect quality*" (ANSI/ASQC Q91, paragraph 4.9.1 (a)).

Based on the above references and citations, one does not need to embark on a mammoth documentation effort requiring (as I have been told by some) as much as two man years. The ISO committee, or the person in charge of addressing paragraph 4.9, Process Control, should focus on the key processes that may affect product quality. Moreover, references to technical standards/codes as they apply to a particular process can also be used or referenced.

(3) *"How many procedures do I need to write?"*

As many as you think is necessary. One is probably not enough, and a thousand is likely to be too much and certainly too expensive. I do realize that this is not a very helpful answer but there is no right answer. If you could afford to document everything and write procedures for every process you will probably find out that if you were to follow your plan to the letter, activities at your plant would come to a grinding halt. To prevent that from happening more and more waivers or exceptions would have to be issued. On the other hand, if you did not have any procedures whatsoever, anarchy would reign. Anyone could do what he/she pleases. Both approaches can soon lead to economic failure. A compromise must be found.

(4) *"Since everything affects quality, must I document everything?"*

Some people would argue that in a broad sense, everything indeed can affect the quality of a product. If a pump leaks and its seals are not promptly replaced, off-spec product could be produced. Similarly, if an operator is not properly trained to read a particular set of gauges, defective product might be manufactured.

Although few would find faults with the above arguments, the fact remains that if you begin your implementation efforts by stating that everything affects the *final* quality of the product you manufacture, you will probably never "get-off the ground." Indeed, the argument that everything affects quality is occasionally proposed by people who do not see any advantage in implementing an ISO type quality system. The logic often put forth is that "it would take at least three man years to document all of my maintenance procedures and I don't have time as it is to do the job properly!" The fact that people do not have the time to do more than what their current job allows for is certainly a reality heard all over the world (the author has heard similar statements in the U.K., France, Norway, Spain and Germany). Fortunately not everything needs to be documented. Also, what needs to be done can be spread over several weeks by a team of individuals sharing the responsibilities.

(5) "I am currently writing all of my maintenance procedures (several hundreds of them), but I can tell you right now that my guys won't read them!"

This actual statement made by a maintenance supervisor is puzzling. Are people not reading the procedures because they know them by heart or because they don't believe in them? If they don't believe in them, then surely the supervisor is wasting a lot of time. Perhaps he should interview his operators to find out how "they do it." If the operators' experience is such that they know the procedures (and in fact have helped write them), then there should not be much concern. Finally, although ISO does not specifically call for the writing of maintenance procedures, other standards such as OSHA for example, might well require more detailed documentation.

(6) "We co-own our processes, how do you handle that?"

I don't really know except to suggest that both parties should work together on the required documentation.

(7) "I have over 5,000 instruments, gauges and other pieces of equipment. Must I calibrate and do precision studies on all of them?"

Certainly not. This question relates to paragraphs 4.11 of ANSI/ASQC Q91 and 4.10 of ANSI/ASQC Q92 (both paragraphs are identical). The opening sentence of 4.11 reads as follows:

> *The supplier shall control, calibrate, and maintain inspection, measuring, and test equipment, whether owned by the*

supplier, on loan, or provided by the purchaser, ***to demonstrate the conformance of product to the specified requirements*** [ANSI/ASQC Q91-1987, paragraph 4.11].

The paragraph continues for another ten subsections (a-j). The most important question in that paragraph is, "do we need to control, calibrate, and maintain inspection, measuring, and test equipment in order to demonstrate the conformance of product to the specified requirements?" Probably not. Although many plants do keep calibration and maintenance records on just about every piece of equipment, not ALL pieces of equipment are routinely calibrated or otherwise tested simply because doing so would be prohibitively expensive. In most cases, equipment such as on-line analyzers or thermocouples for example, are tested on an as needed or as requested basis (see nonetheless the third paragraph of 11.3 in ANSI/ASQC Q94-1987 regarding preventive maintenance). This is certainly not the case for laboratory equipment. Here, a much more rigorous program of calibration, accuracy and precision assessment (perhaps via Statistical Process Control), document control and labelling is certainly required because the laboratory usually is the organization involved in the testing of product specifications directly related to the acceptability and eventual final quality of the product(s).

The rigor required to interpret section 4.11 depends on the type of industry. The author was once told by the quality assurance manager of a pharmaceutical firm that her company had a rigorous instrumentation inspection program on approximately 3,000 pieces of equipment. The person sitting next to her could only think of approximately half a dozen scales which he thought were regularly calibrated by an external agency. It all depends on what is important to you and your customers.

One way to address paragraphs 4.10 (Q92) or 4.11 (Q91) — Inspection, Measuring, and Test Equipment — is to subdivide all pieces of equipment into two or three broad categories: laboratory, process analyzer/instruments and all others. Even after you have segregated your instrument the standards are still relatively demanding. With regard to laboratory equipment, you have little alternative but to abide by ALL of the sub-paragraphs. As for the other types of equipment the guidelines are somewhat more lenient in that paragraph 13.1 (Measurement Control) of the *Guidelines* ISO 9004/Q94 states:

> ***Sufficient*** *control should be maintained over* ***all measurement systems*** *used in the development, manufacture, installation, and servicing of a product to provide confidence in decisions or actions based on measurement data. Control should be exercised over gauges, instruments, sensors, special test equipment, and related computer software* [ANSI/ASQC Q94, paragraph 13.1].

Still, no matter how relaxed the interpretation of the words "sufficient control," the task could be monumental. One may very well wonder, as I often have, why hundreds or even thousands of gauges/sensors are installed in a plant, especially chemical plants, and yet few people seem to know if or when these instruments were calibrated or how accurate they are. I have often been told that despite the hundreds of gauges/sensors available throughout a plant, operators and process engineers only rely on a few key instrument readings to reach a decision regarding the state of a process. If that is indeed the case, you should identify these key measurement points and ensure that your maintenance department or some outside agency, regularly monitors those particular instruments. Be sure to rationalize either in your tier one or tier two documentation, how and why you have sub-grouped the instruments/gauges/sensors in a particular way. Still, one would like to know what is the purpose of these hundreds of other gauges if they are rarely read or monitored!

(8) "My operators know what they have to do because they have done it for fifteen or more years. Do I really need to write procedures?"

Both standards (9001/Q91 and 9002/Q92) recognize the value of job experience. Paragraphs 4.17 (Q92) and 4.18 (Q91) state that "Personnel performing specific assigned tasks shall be qualified on the basis of appropriate education, training, **and/or experience,** as required." Nonetheless, you still will need to prove, via records, that John Doe has indeed been on the job for fifteen years and that he was trained by Mike Knowall back in 19??. Training needs for all personnel would still have to be addressed. Remember that OSHA training requirements are even more demanding.

(9) "How do you address corrective actions when they depend on the type of problem?"

It is certainly true that the same type of problem might be caused by different factors. Similarly, on a particular process, a dozen or more corrective actions might be taken every week. But surely, the Pareto principle must operate as it always does. Among the many corrective actions that are undertaken every day, the same few probably do keep recurring. Do you keep records of these corrective actions? Do you analyze your records to detect patterns and take action to prevent them from recurring? Do Quality Action Teams (QAT) apply their problem solving skills? Are records kept of their (QAT) meetings? Are action items resolved within a certain time period? If not, why not? Answering these questions will help you write your corrective action tier one and tier two procedures.

Experts in troubleshooting, defined as the search for the hidden cause or causes that lead to inadequate performance, perform tasks that are essentially identical to problem solving teams. Consequently, their expertise can be of value to anyone involved with problem solving. In an article published in the April, 1991 issue of *Chemical Engineering Progress*, entitled "Systematize Troubleshooting Techniques," Manfred

Gans, D. Kohan and Burt Palmer offer some insightful comments regarding troubleshooting. Their suggestions for developing good troubleshooting skills closely parallel techniques used in problem solving.

Maintenance trouble, the authors tell us, is caused by five types of failure:

- misoperation,
- false alarm,
- equipment breakdown,
- inadequate equipment design,
- process failure.

As is the case with most problems, we learn that "Big failures are, by and large, due to simple causes, while marginal failures are due to complicated causes." It is interesting to note that the Pareto principle also applies to trouble shooting: "5% of all trouble is due to process failures, 20% is due to inadequate design, and 75% of all unidentified, inefficient plant performance is eventually traced to simple equipment breakdowns." Finally, the authors wisely conclude by stating: "Awareness of interdependence is the foremost skill that must be developed to become a proficient troubleshooter."

(10) "Purchasing is not part of our plant. What do I need to do?"

Although purchasing is not part of your plant you may have to include it as an internal supplier/customer. The third party audit team can (and will likely) request to audit the purchasing department. If the purchasing department is located in another city or state, a separate visit will be required.

(11) "Some of my customers have asked me to document my costs of quality. Would I have to do the same for ISO certification?"

I am often told by process engineers that they invariably have to accept off-spec materials because "we have to run on schedule." When I ask these engineers if they have to adjust their process to compensate for the off-spec material, the answer is invariably, "of course, we have to!" If I then ask the same engineers, "do you monitor and record your internal cost of process adjustments?" the answer is all too often "no." Yet, would it not be useful to know how much off-spec materials cost a plant? Shouldn't a company *measure* its suppliers not only in terms of on time delivery and quantity delivered but also in terms of how much off-spec materials increase its operating cost?

If you don't already do so, the ISO standards will soon require you to measure all of your costs of poor quality. Indeed, the cost of quality is mentioned in paragraphs 0.4.3.1 and 0.4.3.2 of the ANSI/ASQC Q94 *Guidelines* and the next updates of the ISO standards (due sometime in 1992) will include references to the internal and external costs of quality.

(12) "How do I determine what parts of my processes need third tier documentation?"

This question is very much related to questions (3) and (4). A simple and yet efficient way to determine where third tier documentation is needed along one or more processes is to ask the following question: "At what points along the process do we sample or otherwise monitor the process for product quality?" Figure 9.1 below will help illustrate the point.

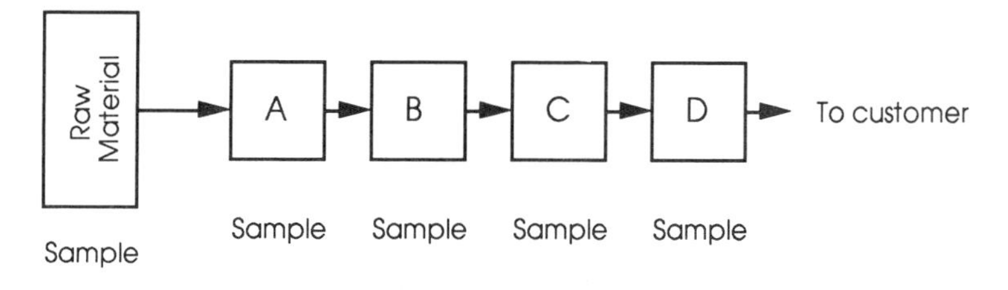

Figure 9.1 A Simple Process

The above schematic representation illustrates a process which consists of four sub-processes. If samples are taken at the indicated sample sites, there is very good reason to assume that these sites are key process check points. Consequently, good sampling documentation AND standard operating procedures should be available at or near these sites. For example if a process requires close monitoring of pH and viscosity, third tier documentation must be made available regarding the operating procedures which ensure pH and viscosity controls.

(13) "Will ISO certification help reduce customer audits?"

Probably not in the short run (i.e. next eighteen to twenty-four months). Although customer audits often parallel or even duplicate the ISO standards, customers will still want to verify if your processes run at a Cpk of 1.33 or better for example. ISO does not yet (and perhaps never will) verify your processes at such a level. One would hope that in the years to come, as customers and suppliers alike begin to appreciate the advantages of certification, customer audits will become less intensive and focus only on specialized portions of the process.

Conclusions

It is impossible to anticipate every question. Recently, I was asked by a seminar participant if his company would have to require all of its suppliers to achieve ISO registration? Certainly not (see Q91 clause 4.6.2). Nevertheless, some U.S. multinationals are currently requiring their suppliers to achieve ISO 9000 registration. Such a policy invariably leads to other important questions such as: "How do you expect small shops (less than ten employees, for example) to be able to afford the cost of ISO 9000 registration? Won't they run the risk of going out of business?" Probably not, since the cost of achieving registration is proportional to the size of an organization and should therefore be within the reach of any company small or large (see previous chapter). Some organizations have already anticipated the need to assist small companies with their total quality efforts including ISO 9000 registration. Consortia of small companies are being formed throughout the nation, particularly in Texas, Arkansas, Indiana and other states to help reduce training and implementation costs. Individuals interested in such options can contact the nearest chamber of commerce or the author for further information.

The above set of thirteen questions are but a subset of the myriad of questions asked by potential users of the ISO 9000 series. I hope the analysis presented herewith will help the reader get a better feel for how to answer other questions he/she might have.

Having spent a considerable amount of time analyzing the standards, it is now time to shift emphasis and focus our attention on another part of the registration process namely, the registrars.

11 On Registrars and EN 45011/2

Once the quality assurance system has been documented, an accredited registrar must be consulted to conduct the third party audit. Which registrar should be considered and what type of service can be expected are two often asked questions.

Not all Registrars are Created Equal

Not anyone can issue ISO 9000 certificates of conformance. To do so, the certification body must itself be accredited by an official agency. Quality directors looking for a registrar must recognize that although over a dozen registrars can issue certificates, *not all certificates are recognized by an official government agency.* Consequently, one must be ***very*** careful when selecting a registrar. The important question to ask is: "Is this certificate recognized in France, Belgium, England or all of them?" For example, if your firm exports mostly to the United Kingdom, you will want to know if the registrar (and hence the certificate) is accredited with the National Accreditation Council for Certification Bodies (NACCB) (3 Birdcage Walk, London, SW1H9JH, England). In addition and most importantly, you will also need to know if your certificate will also be issued with the Crown Stamp of the NACCB. As far as the NACCB is concerned, certificates issued only with the registrar stamp (each registrar has its own distinct stamp) are not valid. The NACCB publishes a *Directory of Accredited Certification Bodies* which is **NOT** to be confused with the *Association of Certification Bodies* published by British Standards Institution (2 Park Street, London, W1A2BS). This latter publication lists all certification bodies (accredited and not accredited).

Until December of 1991 no one really knew which organization — the ASQC's Registration Accreditation Board (RAB) the American National Standards Institute (ANSI) or others ? — was going to be the "national" accreditation body of the United States. Finally, on December 13, 1991, a National Program for Quality Registrar Accreditation was launched by a joint ANSI/RAB program. The joint program is known as the American National Accreditation Program for Registrars of Quality Systems. The December 13 press release announced that "the RAB will administer the overall operations of the program. ANSI is the coordinator of the U.S. private-sector administered voluntary standards system and the official U.S. member body to the ISO and the International Electrotechnical Commission (IEC), via its U.S. National Committee. In this capacity, ANSI coordinates the joint program's due process procedures and provides the program with national recognition *and international acceptance.*"

Which Registrar to Choose?

Although you can select any accredited registrar, you *might* have to ensure that your facilities fit within the *scope* of the registrar. I have emphasized *might* because there are currently two very different philosophies as to whether or not a registrar should or should not have a clearly defined scope. The scope helps you determine whether the registrar will have the required competence/knowledge of your industry. The NACCB and the Institute of Quality Assurance (London) firmly believe that a registrar should not be allowed to audit a particular industry unless it has certain competency within that industry. As was recently explained to me by the director of the NACCB, "auditors having experience in the shoe industry should not be allowed to audit the aerospace industry." The scope of each accredited registrar is listed in the *Directory of Accredited Certification Bodies*. Not all registrars have the same scope, some are more limited than others. In most cases, if competency cannot be found within its ranks, a registrar will try to contact a knowledgeable auditor by referring to a register of certified auditors. Currently, the demand for audits far exceeds the availability of auditors. Consequently, although a registrar will always have certified auditors perform an audit, it cannot always guarantee that the auditors will be the most knowledgeable in the particular industry being audited. This difficulty, which cannot be blamed on the registrars, can lead to some awkward moments during audits. Indeed, since the ISO 9000 series is generic in its content, auditors must interpret and adapt each paragraph to the particular industry they are auditing. The natural tendency for most auditors is to rely on their previous experience. Although this can be helpful in most cases, there are times when processes are very different. For example, a chemical plant operating half a dozen continuous processes cannot be compared to a small or medium sized shop manufacturing a few thousand components a week!

Allan J. Sayle, in his excellent book *Management Audits*, expresses similar concerns when he writes,

> In the ISO 9000 series, for example, one phrase that crops up time and again is "establish and maintain." The auditor must at all costs avoid reading his own experience into this phrase when interpreting the standard. A number of different systems might satisfy the standard's words. Some systems may be cheaper than others and some may appear better than others but seldom will a standard dictate the detail of any specific type of system. All they require is that some particular activity is performed. Two companies with very different systems may satisfy the standards equally well. This fact begins to indicate the risks of relying on assessments performed by third parties who cannot interpret a standard for a particular customer's needs. The more imprecise or weaker the standard, the lower the value of a third party assessment and its resultant certificate. [Allan J. Sayle, *Management Audits*, p. 6-2].

Others do not share the NACCB's view and indeed have vociferously questioned whether it should be the *only* point of view. As far as the European Network for Quality System Assessment and Certification

(EQNET) is concerned (a network of third-party certification bodies), there are only four broad categories of competence (i.e., scope): hardware, software, processed materials and service. The argument proposed by the EQNET is that in order to audit a quality system, one need not be competent in all of the technical aspects present "behind" the quality system. According to this philosophy, experts in the service industry should be allowed to audit any service industry. This certainly seems to be the philosophy ascribed to by the ISO/TC 176 committee.

In an article published in the May 1991 issue of *Quality Progress* entitled "Vision 2000: The Strategy for the ISO 9000 Series Standards in the '90s," the authors — all members of the Ad Hoc Task Force of the International Organization for Standardization Technical Committee 176 — suggest that:

> *Auditors should be accredited (certified)* ***generically,*** *not on an industry/economic-sector basis. Each audit team should include at least one person knowledgeable in the industry/economic sector(s) involved in a particular audit. This knowledge might reside in the accredited auditors on the team or in technical experts on the audit team [Quality Progress*, May, 1991, p. 31].

Having worked with auditors ascribing to both philosophies, I have noted their idiosyncrasies. Some expert auditors have a propensity to forget that they are auditing a quality system. As experts knowledgeable in a particular industry, they occasionally have a tendency to revert back to their second-party role and offer advice and recommendations on how to best fix a problem rather than focus on the system as a whole. Auditors with experience in product (i.e. hardware) testing for example, have on occasion had some difficulties evaluating a quality system, particularly if it is from an industry they are not familiar with. Although both approaches can defend their own rationale, both are faced with potential difficulties which time and experience will no doubt mitigate.

U.S., British and Other Registrars

There is a belief among some quality managers that a firm can only be registered to one of the ISO standards either by a British registrar, an associate of a British registrar or a registrar accredited by the U.K.'s NACCB. Is that necessarily true? Although one cannot deny that the U.K.'s Institute of Quality Assurance (IQA) and the NACCB have taken a lead in organizing and implementing a sound accreditation program, one should not assume that the British accreditation scheme is the sole legitimate program. Besides the NACCB, one must also recognize Germany's TUVCERT in Bonn and the Deutch accrediting body Raad voor de Certificatie, which has accredited American (and foreign) registrars such as Houston based ABS Quality Evaluation, Inc. and Virginia based Intertek. Registrars from Norway (DnV), France (Bureau Veritas), Canada as well as Germany (TUV Rheinland) and other European countries (e.g. Belgium's AIB Vincotte), as well as other U.S. registrars are also available for consultation (see Appendix B for a list of registrars).

The important issues that need to be considered when selecting a registrar have been presented in the previous two sections. The only additional points to consider is that some registrars are accredited by more than one government agency. Some of the foreign registrars currently operating in the U.S. are accredited by the U.K.'s NACCB and the Deutch registration board (R.v.C). Registrars who have been in existence for several decades (some, for over 120 years) are even recognized in four, five or more countries. As for U.S. registrars, two groups exists: those accredited by Milwaukee RAB and those accredited by the RAB as well as the Deutch accreditation board (one American registrar appears to be accredited with the Deutch government but not the RAB). To the author's knowledge, only one U.S. registrar is accredited by the NACCB (this may change by the time this book is published). This does not mean that American registrars have failed to pass the NACCB's accreditation scheme but rather that the NACCB will not recognize a registrar unless it is represented in the U.K.

As far as bilateral or multilateral recognition among European registrars is concerned, the EC has, to my knowledge, not yet (December 1991) released an official list of European registrars. [Note: The author was shown during one of his trips to Europe an *unofficial* list which included the European registrars listed in Appendix B as well as some from Italy, Spain, and other EC member countries.]

One must finally realize that the selection of a third-party registrar is not *imposed* by any customer. The selection process depends on your own criteria and concerns about certification issues. My advice to anyone "shopping" for a registrar is to adopt the "insurance principle" and get at least three opinions.

EN 45011/2: Criteria for Certification Bodies

Rules and accreditation regulations to certificate *quality systems* are set out in the European Standard EN 45012 "General Criteria for Certification Bodies Operating Quality Certification." Criteria for bodies certificating *products* are specified in EN 45011, "General Criteria for Bodies Operating Product Certification." A summary of the nineteen clauses set in EN 45012 are listed below to inform potential customers as to their rights. The set of *Guidelines on the Application of EN 45011/2 to Certification Bodies Operating* can be purchased from the NACCB (see Appendix B).

CLAUSE 1 Object and field of application

This European Standard specifies general criteria that a Certification Body operating a quality system certification shall follow if it is to be recognized at a national or European level as competent and reliable in the operation of quality system certification, irrespective of the sector involved. It is intended for the use of bodies concerned with recognizing the competence of certification bodies.

CLAUSE 2 Definitions

2.1 *certification of conformity*: Action by a third party, demonstrating that adequate confidence is provided that a duly identified product, process or service is in conformity with a specific standard or other normative document.

CLAUSE 3 General requirements

All suppliers shall have access to the services of the Certification Body. There shall not be undue financial or other conditions. The procedure under which the body operates shall be administered in a non-discriminatory manner.

CLAUSE 4 Administrative structure

The Certification Body shall be impartial and shall have:

(a) a structure which requires the choosing of members of its governing board from among those interests involved in the process of certification without any single interest predominating...

(b) permanent personnel under the senior executive responsible to the governing board to carry out the day to day operations in such a way as to be free from control by those who have a direct commercial interest in the products or services concerned.

CLAUSE 5 Terms of reference of governing body

The governing board shall be responsible for performance of certification as defined by this standard. Its functions shall cover:

(a) the formulation of policy matters relating to the operation of the Certification Body;

(b) an overview of the implementation of its policies;

(c) an overview of the finances of the Certification Body;

(d) the setting up of committees as required, to which defined activities are delegated.

CLAUSE 6 Organizational structure

The Certification Body shall have and make available on request:

(a) an organizational chart showing clearly the responsibility and reporting structure of the organization and in particular the relationship between the assessment and certification functions;

(b) a description of the means by which the organization obtains financial support;

(c) a documented statement of its certification systems including its rules and procedures for granting certification;

(d) documentation clearly identifying its legal status.

CLAUSE 7 Certification personnel

The personnel of the Certification Body shall be competent for the functions they undertake.

Information on the relevant qualifications, training and experience of each member of the personnel shall be maintained by the Certification Body. Records of training and experience shall be kept up-to-date.

Personnel shall have available to them clear documented instructions pertaining to their duties and responsibilities. These instructions shall be maintained up-to-date.

CLAUSE 8 Documentation and change control

The Certification Body shall maintain a system for the control of all documentation relating to the certification system and shall ensure that:

(a) the current issues of the appropriate documentation are available at all relevant locations;

(b) all changes of documents or amendments to documents are covered by the correct authorization and processed in a manner which will ensure direct and speedy action at the effective point;

(c) superseded documents are removed from use throughout the organization and its agencies;

(d) certificated suppliers and other users of its certification schemes are notified of changes. This may be accompanied by direct mailing or by issue of a periodic publication.

CLAUSE 9 Records

The Certification Body shall maintain a record system to suit its particular circumstances and to comply with any existing regulations.

The records shall demonstrate the way in which each certification procedure was applied including assessment and surveillance.

All records shall be safely stored for an adequate period, held secure and in confidence to the client, unless otherwise required by law.

CLAUSE 10 Certification and surveillance procedures

10.1 The Certification Body shall have the required facilities and documented procedures to enable the assessment, certification and surveillance of quality systems to be carried out in accordance with the requirements of the relevant international documentation.

10.2 The Certification Body shall require the supplier to have a documented Quality System.

10.3 The Certification Body shall maintain regular surveillance of the supplier's quality system.

CLAUSE 11 Certification and surveillance facilities required

11.1 The Certification Body shall have the required facilities in terms of certification personnel, expertise and equipment to perform assessment, certification and surveillance of the supplier's Quality System for compliance with the requirements. This does not preclude the use of external resources when necessary.

11.2 If assessment or surveillance is carried out on behalf of the Certification Body by the external body, the Certification Body shall ensure that this body conforms to the requirements of 11.1. A properly documented agreement covering these arrangements including confidentiality shall be drawn up.

CLAUSE 12 Quality Manual

The certification Body shall have a Quality Manual and documented procedures setting out the way in which it complies with the criteria.

The information supplied shall include at least:

(a) a quality policy statement;

(b) brief description of the legal status of the Certification Body;

(c) a statement of the organization of the Certification Body, including details of the governing board, its constitution, terms of reference and rules of procedures;

(d) names, qualifications, experience and terms of reference of the senior executive and other certification personnel, both internal and external;

(e) details of training arrangements for certification personnel;

(f) an organization chart showing lines of authority, responsibility and allocation of functions stemming from the senior executive;

(g) details of the documented procedures for assessing and auditing supplier Quality Systems;

(h) details of documented procedures for surveillance of suppliers;

(i) a list of sub-contractors and details of the documented procedures for assessing and monitoring their competence;

(j) details of appeals procedures.

CLAUSE 13 Confidentiality

The Certification Body shall have adequate arrangements to ensure confidentiality of the information obtained in the course of its certification activities at all levels of its organization, including committees.

CLAUSE 14 Publications

14.1 The Certification Body shall produce and update as necessary a list of certificated suppliers with an outline of the scope of the certification of each supplier. The list shall be available to the public.

14.2 A description of the certification systems shall be available in published form.

CLAUSE 15 Appeals

The Certification Body shall have procedures for the consideration of appeals against its decisions.

CLAUSE 16 Audit and periodic review

The Certification Body shall undertake internal audits and periodic reviews of its compliance with the criteria of this standard. Such reviews shall be recorded and be available to persons having a right of access to this information.

CLAUSE 17 Misuse of certificate

17.1 The Certification Body shall exercise proper control on the use of its Quality System certificates.

17.2 Incorrect references to the certification systems or misleading use of certificates found in advertisements, catalogues, etc. shall be dealt with by suitable actions.

CLAUSE 18 Complaints

The Certification Body shall require the certificated suppliers to keep a record of all complaints and remedial actions relative to the Quality System.

CLAUSE 19 Withdrawal and cancellation of certificates

The Certification Body shall have documented procedures for withdrawal and cancellation of Quality System certificates.

The Registrar Accreditation Board (RAB) has adopted similar guidelines. Anyone interested in obtaining the set of guidelines should contact the RAB at the address listed in Appendix B.

Conclusions

No rules can be set regarding the selection of a registrar. They are all competent. Even though most registrars will operate very similarly, they each have their own "personality" and style. Each has its own format for issuing nonconformance. Some issue minor and major nonconformities, others use a point (rating) system. It is your responsibility to select the registrar that best fits your business needs and corporate profile. Some companies elect Canadian registrars simply because they either export mostly to Canada or because their Canadian plant(s) first contacted a Canadian registrar. Others prefer non-American registrars either because of their long time expertise, overseas recognition, or simply because they mostly export to one particular country. Still others, not wanting to rely on foreign registrars, select one of the three (as of December, 1991) American registrars. The only way to ensure that you have selected the right registrar is to contact more than one. Ask as many questions as possible, find out who they are and how they operate. Then, and only then, will you be able to make a rational decision.

12 The Third Party Audit

> ***Quality Audit:*** *A systematic and independent examination to determine whether quality activities and related results comply with planned arrangements and whether these arrangements are implemented effectively and are suitable to achieve objectives* [ISO 10011-1:1990, 3.1 p. 1].

Introduction

The following account covers much information that is contained in the ISO 10011-1, 2 and 3 series. These three *Guidelines for auditing quality systems* are respectively entitled: *Auditing* (Part 1), *Qualification criteria for quality systems auditors* (Part 2) and, *Management of audit programs* (Part 3). The information contained in these three pamphlets (the longest being seven pages), parallels the subject matter covered in "Lead Assessor" courses. For further information, the reader should consult one of the many agencies offering these "Lead Assessor" courses.

On Audits

Quality audits are *independent* evaluations conducted to compare various aspects of quality performance with a standard for that performance. Quality audits have been used by companies and government agencies to evaluate their own quality performance (first party or *internal audits)*, or the performance of their supplier/vendors/sub-contractors (second party or *external audits*).

Audits conducted in the spirit of a customer-vendor *partnership* can be beneficial in that they provide an excellent opportunity for both parties to "get to know" each other. Quality audits should never be intended to be *punitive,* rather they should be conducted in a spirit of cooperation. Indeed, audits should go beyond finding out how a quality system is conforming to some pre-determined set of standards; they should also allow for continuous improvement. Many supplier quality assurance type audits such as Ford, John Deere, 3M and others have been designed not only to ensure thoroughness but also to motivate a spirit of partnership.

Traditionally, suppliers have been audited by either their customers or a third party to provide *independent* assurance that:

Plans for attaining quality are such that, if followed, the intended quality will, in fact, be attained.

Products are fit for use and safe for the user.

Laws and regulations are being followed.

There is conformance to specifications.

Procedures are adequate *and* are being followed.

The data system provides accurate and adequate information on quality to all concerned.

Deficiencies are identified and corrective action is taken.

Opportunities for improvement are identified and the appropriate personnel alerted.

Types of Audits

There are two types of audits: audits of activities (*systems audit*) and audits of product *(product audit)*. *Product audits* involve the reinspection of product to verify the adequacy of acceptance and rejection decisions made by inspection and test personnel. *Systems audits* include any activity that can affect final product quality. Quality systems audits are directed at the quality aspects of various segments of the overall approach to quality. A typical list of such systems would be:

General administration of the quality function.
Product development and design.
Vendor relations.
Manufacturing planning.
Process control.
Final product testing.
Internal audits.

Systems oriented audits can review any or *all* of the above systems over a whole range of products. As a rule, systems audits are more time consuming than product audits. Although the time spent auditing a product or a system depends on the complexity of the product and the size of an organization, most audits last anywhere between two to five days. Third party audits based on the ISO 9001, 9002 or 9003 series of standards are examples of a systems oriented audit.

Second Party Versus Third Party Audits

Second party audits generally known as Supplier Quality Assurance audits are either product or system audits. The purpose of such audits is to determine whether or not the supplier *conforms* to some pre-specified contractual procedure imposed by the buyer/customer. When non-conformities are found, the *burden of proof is on the auditee.* Third party (independent) audits have a different philosophy. With third party auditors, the burden of proof regarding nonconformities *is on the auditor! Thus before raising a non-conformance, the third party auditor must find objective evidence (i.e., proof) that what is claimed does not match what is done.* Objective evidence is defined as follows:

Definitions

Observation: A statement of fact made as part of the audit process and substantiated by *objective evidence.*

Objective Evidence: Qualitative or quantitative information, records or statements of fact pertaining to the quality of an item or service or to the existence and implementation of a quality system element or documented requirement that is based upon observation, measurement or test and which can be verified [ISO 10011-1, p. 2].

Third Party Auditors

Although all third party auditors are certified as having passed a standardized lead assessor training course and have conducted at least three to five audits prior to participating in an actual audit, problems do arise. The complex issue of "calibrating auditors," has been on the author's mind for quite some time. Having trained auditors, I can attest to the fact that despite training, not all auditors will interpret a standard similarly. This is to be expected since the interpretation of a standard is not only a function of training but also of the auditor's psychological profile. Some auditors are flexible and understand how to interpret the generic nature of the ISO standard, others rigidly interpret every paragraph. The auditor's background, and industry experience are also important factors to consider when selecting a team of auditors. Fortunately, the lead auditor will always be present to provide his expertise (see Chapter 13 for further comments).

Qualification Criteria for Quality Systems Auditors

The qualifications required of an auditor vary according to the registrar and the regulatory body which accredits the registrars. The ISO 10011-2 recognizes seven major criteria:

(1) Education: The auditor "should have completed at least secondary education... (and) should have demonstrated competence in clearly

and fluently expressing concepts and ideas orally and in writing in their officially recognized language" [ISO 10011-2, p. 1].

(2) Training: The auditor should have received training "to the extent necessary to ensure their competence" in carrying out and managing audits. Such training is usually accomplished by attending a five day course recognized by London's Institute of Quality Assurance (IQA). These courses are currently being offered throughout the United States.

(3) Experience: A minimum of four years, plus other requirements are called for by the standard. One wonders why four years would be needed to acquire sufficient competency in the audit process?

(4) Personal attributes: Open-mindedness, maturity, sound judgment, analytical skills, tenacity and an ability to understand complex operations "from a broad perspective" are some of the attributes specified by the standard.

(5) Management capabilities: Auditors should be able to demonstrate their ability to execute all of the steps required of an audit as specified in 10011-1 (e.g., planning the audit, documenting observations, communicating and clarifying audit requirements, reporting audit results, etc.).

(6) Maintenance of competence: The auditor should continuously update his knowledge and skill of the audit process.

(7) Language: The auditor must demonstrate fluency in the "agreed language of the audit."

Lead auditors must possess all of the above skills and must in addition, "have acted" as a qualified auditor "in at least three audits performed in accordance with the recommendations given in ISO 10011-1" [ISO 10011-2, p. 2 paragraph 11]. These requirements are minimum requirements. Most registrars require their auditors to participate in at least five audits before being recognized as auditors and five more as "lead auditor trainee," before being given the title of lead auditor (or assessor). [Note: In the U.S., the term assessor is often preferred to auditor. Auditing is an activity often associated with the Internal Revenue Service; a rather notorious government agency.]

The above ISO 1001 requirements are but a generic model. National Registration Schemes usually append additional requirements or modifications to the ISO model. The U.K.'s NACCB for example, has a scoring system for evaluating qualifications and experience. The Registrar Accreditation Board of Milwaukee modeled its requirements on the NACCB and ISO guidelines. The RAB requires auditors to have an impressive (but questionable) ten years experience! Fortunately, the experience can be achieved via a combination of education, training and on the job experience.

The Third Party Audit Process

Although the audit process will vary slightly from registrar to registrar, the pattern pretty much follows guidelines set in ISO 10011-1 and 10011-2. From the prospective client's point of view, one of the most important facts to retain concerns the *Audit scope.* As paragraph 5.1.1 of ISO 10011-1 explains:

> The client makes the final decisions on which quality system elements, physical locations and organizational activities are to be audited within a specified time frame. This should be done with the assistance of the lead auditor [ISO 10011-1, p. 4].

Having defined the scope the client should next provide sufficient objective evidence to demonstrate that his quality system functions effectively. Assessment of the quality system's effectiveness is the primary objective of the third party review. This activity consists of several steps which are:

1. *Pre-audit Visit*

The purpose of the pre-audit visit is many fold. Basically the visit allows the lead assessor to get to know the client and familiarize himself with the site location (e.g., plant layout), so as to better determine how many man-days will be required and what expertise he will need on his team on the day of the audit. The pre-assessment visit also helps the lead auditor determine the client's readiness vis-a-vis the ISO standard.

2. *Documentation Review or Adequacy Review*

Having determined that you satisfy all of the pre-requirements, the accrediting agency will next need to review a copy of the quality manual to see if it conforms to all of the pertinent ISO (9001, 9002 or 9003) requirements.

3. *Preparation Audit Trail Matrix (Compliance Audit)*

Next, the lead auditor will prepare (with the assistance of his auditors) the *audit plan.* This usually is nothing more than an audit trail matrix which identifies who will audit what (see the audit trail matrix on [age 153). It is interesting to observe that according to paragraph 5.2.1 of ISO 10011-1, "The audit plan should be approved by the client and communicated to the auditors and auditee." The author has not been able to verify if ALL registrars will allow their clients to review the registrar's audit plan.

4. *Establish Program*

Send audit program 7-28 days prior to audit along with résumé of auditors. Program will include such items as:

Facilities requirements.
Escorts.
Meals + accommodations.
Attendance at opening and closing meetings.

Sample Audit
Matrix for Three Auditors

Check marks are for illustration purpose only

ISO Quality System Requirement Section	Purchasing J.L.L.	Marketing S.E.	R&D J.L.L.	Servicing S.E.	Packaging R.W.		• • • • • Other Functional Departments • • •
4.1	✓	✓	✓	✓	✓		
4.2	✓						
4.3					✓		
4.4	✓			✓			
4.5				✓	✓		
4.6				✓			
4.7				✓		✓	
4.8					✓		
4.9		✓			✓		
4.10				✓			
4.11	✓						
4.12				✓			
4.13		✓					
4.14						✓	
4.15	✓		✓			✓	
4.16			✓				
4.17			✓				
4.18	✓					✓	

[Note: You have the right to review each auditor's credentials. Having reviewed each resume you may then accept or reject anyone.]

5. *The Audit*

Although audit procedures may vary slightly, the following agenda is generally considered standard procedure:

8:00 A.M. Team briefing.

9:00 A.M. Opening Meeting (15 minutes)

- Summary of program.
- Review the audit's scope.
- Brief description of possible outcomes:

  1. Recommendation
  2. Deferred decision
  3. Termination of audit

- Arrangements to communicate closing meeting (office space).
- Start audit.

If nonconformities are found (raised), the auditor must turn in evidence (observations) to the lead auditor, who in turn adjudicates. With regard to the collection of nonconformities ISO 10011-1 paragraph 5.3.2.1 states:

> Evidence should be collected through interviews, examination of documents, and observations of activities and conditions in the areas of concern. Clues suggesting nonconformities should be noted if they seem significant, even though not covered by check-lists, and should be investigated. Information gathered through interviews should be tested by acquiring the same information from

other independent sources, such as physical observation, measurements and records" [ISO 10011-1 *op. cit.* p. 5].

All final nonconformities must be included in a report which will include: (1) the scope of the audit, (2) details of the audit, such as participants' names, date of the audit, (3) list of nonconformities with ISO reference, (4) the team's assessment as to the client's quality system compliance and (5) a distribution list.

6. *Closing Meeting (One Hour)*

The purpose of the meeting is to:

- Present all nonconformities.
- Offers recommendations (not necessarily suggestions).
- Provide individual comments from each auditor.
- Present concluding remarks.

If nonconformities are raised, as they are likely to be, there is no need to panic. Corrective actions can be submitted generally within 30 days. If more days are needed, this can be negotiated with the lead auditor. When all corrective actions have been addressed the lead auditor will determine whether or not another visit will be required. Once recommended for certification, as much as 45-60 days may elapse before reception of the certificate with the appropriate stamps.

7. *Surveillance*

The quality system will be periodically re-visited. The frequency of these visits will depend on the registrar. A partial audit every six months for some, a complete audit every three years for others. If serious nonconformities are identified during surveillance, certification can (and will) be revoked.

Conclusions

Third party audits or assessments should not be an unpleasant experience. Although they will not necessarily be delightful experiences, they will always be informative. The auditee must recognize that the registrar's objective is not to fail anyone but rather to verify that the quality system conforms to the ISO standard. If the quality assurance system is not in conformance, the lead auditor will point out what needs to be addressed in order to align the company's quality assurance system with the standard. It is perhaps comforting to know that approximately seventy to seventy-five percent of all companies that have gone as far as their first ISO audit will eventually achieve registration.

One of the best ways to prepare for a third party audit is to go through a first (i.e., internal) audit process. Having reviewed the fundamentals of third party audits, we next offer suggestions on how to facilitate internal audits.

13 How to Facilitate an Internal Audit

The Need for Internal Audit

The conduct of internal audits is clearly specified by ISO (paragraph 4.17 of ISO 9001/Q91 and 4.16 of ISO 9002/Q92). Moreover, paragraph 4.1.2.2 entitled *Verification Resources and Personnel*, specifies that "The supplier shall identify in-house verification requirements, provide adequate resources, and assign trained personnel for verification activities." To ensure compliance with paragraphs 4.17 and 4.1.2.2, many companies enroll several of their employees into one of the many five day lead assessor courses currently offered by the few "officially approved" agencies. Some multinationals have even invited consulting agencies for in-house two day auditor training for as many as twenty to sixty of their employees.

A Word of Caution

There seems to be some confusion and occasional unrealistic expectations as to what a five day lead assessor course does provide. First of all, the course must be an approved course which means that its contents (including case studies) have been reviewed and approved as being satisfactory by some official (i.e., governmental) agency. At present (December, 1991) only one such organization is in existence namely, the Institute of Quality Assurance (IQA) of London, England (the Netherlands might also have a similar organization). The advantage of attending an IQA approved course is that the contents are, more or less, standardized. The author has attended two such courses from two different organizations and can attest to the fact that both courses and case studies were very nearly equivalent in content (although perhaps not in emphasis).

To become an ISO 9000 registered auditor, you must do a bit more than just attend the course. You must also successfully pass a two hour exam (seventy percent of those taking the exam pass it). However, passing the exam *does not automatically qualify you as a registered auditor.* [Note: Some people and consultants seem to think that is all that is required!] This is where the confusion starts. In order to be recognized as an ISO 9000 auditor, you must next participate in three to five audits (each registrar seems to have its own set of rules, although the ISO 10011 series only mention three audits, see comments in previous chapter). These audits must in turn be led by a registered lead auditor. Having conducted the required number of audits, you may then apply (a $75-$100 fee is also assessed) to the IQA for official registration. This in essence means that, if accepted, your name will be included in an official register. Then and only then will you be considered an "official" ISO 9000 auditor. It seems simple enough except when you ask the question: "How do I find out about conducting my first three to five audits?" In other words, how do I join the club? You will soon find out that unless you are willing to work for one of the registrars, you *will (generally speaking) not be able (or allowed) to conduct such audits.* The rules might change as the need for lead auditors increases.

The above procedures do not necessarily apply to the U.S.'s own Registrar Accreditation Board (see Appendix B for address), which has its own auditor certification procedures. Whether the RAB's procedures are recognized by the U.K.'s IQA is not known to the author. Suffice it to say that it is much easier to be recognized as an officially trained (internal) auditor than it is to become a fully accredited ISO 9000 auditor, unless of course you happen to be associated with one of the registrars.

Elements of Successful Interviews

Most audits involve interviewing quite a few people. Since an interview is essentially a method of collecting information, steps must be taken to insure that pertinent information is being gathered.

There are several key factors involved in conducting an interview:

1. No interview should ever be conducted without a plan which has:

   a. An established objective outlining who within the organization will have to be interviewed.

   b. A set time-frame to begin and end.

   c. Allocated time for research of the subject to be covered and the functions of the interviewees.

2. Keep the interview to the subject for which it was planned.

3. Guide the interview with your questions.

4. Listen and be sure that you understand what was said *and meant.* Don't hesitate to cross-check facts with other individuals within the organization.

5. Evaluate the method and the results of the interview.

Preparing for the Interview

Before conducting an audit of your internal supplier/department you should try to secure the following information:

1. An organization (second tier) chart. This will allow you to plan your audit as well as estimate how much time your interviews will take.

2. A schematic of the manufacturing process(es). This can help you formulate specific questions during the audit. Naturally, this step might not be required if you are already familiar with the process.

Conducting the Interviews

Since most audits can last anywhere from two to five days, interviewing anywhere from eight to twelve or more people, it is impossible to remember *who* said *what when*. It is therefore a good practice to always carry along a small handbook whenever touring a plant or interviewing people. For each interview, be sure to include time of day and name of the person being interviewed. If people find it difficult to talk to you while you write notes, try to either limit your note taking to keywords, or simply stop writing altogether.

Whether you take complete or partial notes, you should always allow ten to fifteen minutes at the end of each interview to prepare your own summary of what has been said. Such "in-between interview pauses" are extremely valuable for two reasons:

1. They allow you to (a) collect your thoughts, (b) summarize what has been said which in turns allows you to (c) modify or otherwise prepare your next interview.

2. They will also provide you with a much needed rest period in between interviews. Such pauses are most important if you want to remain an effective listener for the whole duration of the interview. If you fail to schedule these pauses, you will find yourself rushed from one meeting/interview to the next.

Effective listening is not only the key element to any good audit, it is also a surprisingly exhausting task. I would therefore recommend that in order to remain sharp and attentive, the auditor should schedule at least two pauses a day.

Relation Auditor-Auditee

The person being audited can perceive the auditor as:

- A stranger who should not be trusted.

- Someone whom he/she does not really want to talk to but has to.

- Someone who wastes his/her precious time.

- Someone a bit too curious.

- Someone who is looking for a 'crack' in the system ***and who usually finds it***.

In order to avoid mistrust and doubt, the auditor should always clearly state the purpose of his/her audit. This step is equally important for internal audits. The purpose should simply be to verify the effectiveness of the quality assurance system. If the system is ineffective, then recommendations must be made on how to improve the system rather than blindly enforcing it. Punitive actions, negative and particularly destructive criticism or sarcasm must be avoided. Such actions can only lead to catastrophic results.

Formulating Questions

A good question is a question that:

- Helps the auditee express himself.

- Does not accuse or attack.

- Does not influence and allows the auditee to freely answer.

One should not forget that the purpose of any audit is to collect ***facts***; that is why questions must:

- Stimulate thoughtfulness.

- Channel information.

- Provoke a reaction.
- Provide explanations.
- Explore, clarify, verify, illustrate, etc.

The quality audit should be designed to collect *facts* and not vague statements; precision is of the essence.

Answers such as:

"Many," "a few" and "quite a bit" should prompt the auditor to ask "*how much*"?, "always," "never," "often," or "generally," should prompt *"when (does it arrive)?"*

Statement such as:

"Generally"	should prompt	"and in particular."
"In principle"		"and in reality."
"In theory"		"and in practice."

Questions should however avoid referring to *opinions, feelings or vague concepts.*

Types of Questions

There are basically two types of questions: *open-ended questions and closed questions.*

Closed questions are very limiting in that they are conducive to laconic answers such as: "yes," "no," "five," etc. Although these questions lead to a very precise answer, they tend to discourage any dialogue.

Open-ended questions allow for greater flexibility. Examples of open-ended questions would include:

"How do you control bath temperature?"

"Why do you have this particular control?"

"Tell me more about your calibration procedure."

"Could you please explain your incoming material inspection procedure?"

Other Types of Questions

1. Questions designed to obtain clarification.

"What do you mean when you say 'its always the same thing'?"

"Could you give me an example of..."

2. Inappropriate questions.

Inappropriate or false questions are not questions but rather opinions expressed by the auditor. They should be avoided.

"Don't you think that is a lot of rejects (rework, scrap)!?"

"Isn't that a high external cost?"

"Don't you think management is dragging their feet?"

Repeating

It is often very helpful to repeat the answer(s) you have just heard to *confirm* your understanding with the auditee. This mirror effect can be very helpful.

"To summarize, if I understand what you just said,..."

"Therefore, as far as you are concerned,..."

"It is therefore your opinion that..."

Some Advice on How to Ask Questions

1. Avoid asking questions too rapidly. Do not rush through your interview.

2. Avoid asking more than one question at a time.

3. Avoid asking lengthy questions.

Since questions can be a very effective tool to acquire information they must be carefully stated. Do not forget that *from the point of view of the auditee, your questions will generally be perceived as a form of intrusion. Every time you ask a question you in some way impose your frame of reference within which the auditee must try to answer to the best of his/her ability* (see Table 13.1 below for suggestions on how to handle difficult interviews).

Auditor's Behavior

He carefully and attentively listens.

He demonstrates a genuine interest in the explanations provided by the auditee.

He never judges or criticizes.

He is patient.

He neither approves nor disapproves.

He does not prepare the next question while listening to an answer.

He should never be vindictive, nor suspicious but rather trustworthy.

He maintains eye contact.

Table 13.1 De Masi's Suggestions for Difficult Interviews

Respondent Behavior	Interviewer Action
1. Appears to guess at answers rather than admit ignorance.	1. After the interview, cross-check answers that are suspect.
2. Attempts to tell the interviewer what he presumably wants to hear instead of the correct facts.	2. Avoid putting questions in a form that implies the answers. Cross-check answers that are suspect.
3. Gives the interviewer a great deal of irrelevant information or tells stories.	3. In friendly but persistent fashion, bring the discussion back into desired channel.
4. Stops talking if the interviewer takes notes.	4. Put the notebook away.
5. Attempts to rush through the interview.	5. Suggest coming back later.
6. Expresses satisfaction with the way things are done now and wants no change.	6. Encourage him to elaborate on present situation and its virtues.
7. Shows obvious resentment of the interviewer, answers questions guardedly, or appears to be withholding data.	7. Try to get him/her talking about something that interests him.
8. Sabotages the interview by noncooperation.	8. Ask him if you can get the information from someone else.
9. Gripes about his job, his pay, his associates, his supervisors, the unfair treatment he receives.	9. Listen sympathetically and listen for clues. Then suggest how his gripes can be of help to the company.
10. Acts as eager beaver, is enthusiastic about new ideas, gadgets, techniques.	10. Listen for desired facts and valuable leads.

On Silence

It is generally believed that during an interview, twenty to thirty percent of the time should be taken by the auditor and the balance should be granted to the auditee. In order to respect such a ratio, the auditor should learn to remain *silent* however, he should be:

Silent but ***attentive and interested,***

Silent but ***observing and listening,***

Silent but ***taking notes.***

The auditor should always avoid *interrupting the auditee or talking at the same time.*

Where Should I Start?

Unless you are interviewing a particular department, your first meeting should always be with the plant director/manager and his staff which would include the quality manager or his nearest equivalent. Having explained the purpose of your visit, you should then outline your procedure for the next few days. You could either start with *receiving* and end with *shipping* making sure to visit *purchasing, laboratories (test), R&D lab., quality assurance, production schedule, maintenance, packaging, etc...,* or follow the reverse procedure. All along, the guiding principle should be to *always ask for documented evidence of the stated (i.e. verbalized) procedures.*

How Much Time Should I Spend?

Individual interviews should generally last anywhere between thirty to fifty minutes. When terminating an interview the interviewer should always do three things:

1. Give the respondent one last opportunity to mention anything he thinks is significant that hasn't been discussed.

2. Get the respondent's permission to return with any questions the interviewer may subsequently think of.

3. Get the respondent's commitment to review the interviewer's interview report.

On average, for a two to three day audit, you can expect to spend four to eight hours interviewing people, a half day checking documentation and anywhere between five to eight hours touring the plant and/or talking to operators. This leaves little time for the last and most important meeting, debriefing your host.

The Report

The final report should include:

1. A list of the persons interviewed.

2. A summary of the audit outlining the *strong points* and the areas *needing improvement.*

3. If appropriate, the report could include pertinent quotations collected by the auditors during his visit. Such quotations can be extremely valuable when making a point.

4. If a questionnaire was used during the audit, the scored questionnaire should be returned.

5. If the audit is a supplier quality assurance type of audit, the auditor could conclude his report by offering suggestions on how both parties could work together to help resolve problems.

6. A date specifying when and by whom the particular corrective action(s) will be resolved must also be set.

7. Naturally, records of the audit must also be maintained.

8. The audit process itself should be periodically reviewed by management.

How to Be Ready for an Audit? What to Say and How to Say it?

Whenever an organization or department is about to be audited, a certain amount of preparation needs to take place. The following guidelines could be followed:

1. Make sure you know well in advance (three to four weeks notice is considered acceptable) as to when the audit will take place. Some internal auditors prefer to conduct their audits unannounced, perhaps hoping to "catch" the villains in a flagrant act of nonconformance.

2. If applicable, request any documentation (questionnaire or procedures) relating to the audit procedure. Study the document and make sure you are prepared.

3. It is often a good idea to rehearse an audit. In most cases, the pre-audit should be conducted by an outside (independent) source. This will guarantee objectivity. One does not rehearse in the hope of fooling the auditor (although this is often the intent). Rather, rehearsing allows the auditee to focus on the eventual audit (see (4) below).

4. When the auditors arrive, do not try to hide your weaknesses; doing so will encourage a good auditor to be even more inquisitive. On the other hand *do not* volunteer information not requested by the auditor. This can be achieved by keeping your answers *short* and *to the point.* Make sure you have all pertinent documentation ready for the auditors' inspection. Try to show that you have not rehearsed the audit.

5. Finally try not to crowd the auditors. Although it always is a good idea to escort auditors in order to "control" the audit as much as possible, you should also demonstrate enough confidence in your organization by allowing auditors to explore wherever they want and interview whomever they want.

Conclusions

Internal auditing is an important activity that allows departments to continuously improve their function. In the very early stages of implementation, trained internal auditors/assessors should provide valuable assistance and advice on how to prepare for registration. Unfortunately, in some cases, internal audits are either not taken seriously or misunderstood. When internal audits are not taken seriously, usually a sign of limited managerial commitment, corrective actions remain pending. Unable or unwilling to correct the system, departments or department heads wrongly assume that it is the responsibility of the internal audit team or the quality manager to address all corrective actions. This should not be the case. Internal auditors should not be perceived as policeman or enforcer of the "ISO quality assurance system." Their function is to point out deficiencies or inaccuracies within the system. Moreover, since they are not directly responsible for the department being audited (as required by ISO, do you remember which clause?), they cannot nor should not be expected to correct the very nonconformities they raise. These issues are the *direct* responsibility of the non-compliant department.

If the role and function of internal audits are clearly understood by all, much can be gained. Above all, internal audits will *help* you monitor and achieve readiness.

14 The ISO 9000 Series in 1992 and Beyond

Predicting the nature of future upgrades for the ISO 9000 series is riskier than predicting the weather. This is perhaps due to the fact that although weather prediction requires the monitoring of thousands of variables, meteorologists do not have to contend with complex human characteristics such as capriciousness, prejudices and convictions. Indeed, one must recall that the ISO 9000 series of standards must meet with the approval of nearly one hundred members of the international community. Moreover, as is the case with most international forums, various factions interpret the ISO 9000 differently. Consequently, the opinions of one group will not necessarily be shared by others.

What I am about to present are some of the *suggested* revisions agreed by members of Working Group 11. Since these are *suggested revisions* the reader must NOT interpret these comments as having received final *approval* for the upcoming 1992 ISO 9000 series updates. Nonetheless, the proposed revisions do give some general indication as to how the standards *might* be modified.

Most suggested updates could be considered minor updates. This is not to imply that all suggestions are minor but rather that, for the most part, they consist of amendments and expansions or clarifications.

Recommended Updates for the ISO 9000 Series: 1992

Those of us who have had to use 9001, 9002 and 9003 have always wondered why the numbering of the clauses is not identical. Apparently, others have too and an improvement is long overdue.

Whereas it is true that the current (1987) version of the ISO standards does not yet have paragraph(s) on the cost of quality, it is rumored that the 1992 updates will address the very issue described in paragraph 0.4.3, *Cost Consideration*, of ANSI/ASQC Q94-1987.

0.4.3.1 For the Company

Consideration has to be given to costs due to marketing and design deficiencies, including unsatisfactory materials, rework, repair, replacement, reprocessing, loss of production, warranties, and field repair.

0.4.3.2 For the Customer

Consideration has to be given to safety, acquisition cost, operating, maintenance, downtime and repair cost, and possible disposal cost.

In anticipation of these new requirements, companies should begin monitoring their internal failure, external failure, appraisal and prevention costs (for a discussion on quality costs see J.M. Juran and Frank M. Gryna Jr., *Quality Planning and Analysis*, or H. James Harrington's *Poor-Quality Cost*). Many companies are already required to do so by their customers. Perhaps the best approach to consider when tracking poor-quality costs is to assign a team which should consist of a representative from accounting, a process/industrial engineer and a quality engineer.

Safety, mentioned in 9004, will probably be integrated in the standards. Other updates can be broken down into three categories: those calling for an amendment, those requiring a paragraph to be expanded and those falling within the other improvement category.

Clauses Which May be Amended

The following clauses have been selected for various amendments:

Clause	Nature of Amendment
4.2 Quality System	Replace "system" with "planning." Identify requirements for quality manual and quality plans and clarify distinction between requirements for product and for quality system.
4.3 Contract Review	Broaden scope and include timing of pre- and pro-contract stages.
4.4.2 Design and Development	Define responsibility for planned designed activities.
4.4.4 Design Output	Add a clause to ensure that outputs will comply with applicable regulatory and standards requirements even if those have not been stated in input information.
4.11 Inspection, Measuring, etc.	Refer to ISO 10012. Distinguish between types of equipment.
4.17 Internal Quality Audits	Refer to ISO 10011. Also address 2nd and 3rd party external audit and verification criteria.
4.20 Statistical Techniques	Replace "identifying" with "implementing." Give more specific details.

Clauses Which May be Expanded

The following clauses might be expanded:

Clause	Nature of Expansion
4.1.2.2 Verification resources + Personnel	Expand to include all personnel.
4.6.2 Assessment of sub-contractor	Expand text on selection of sub-contractors.
4.6.3 Purchasing Data	Expand to include requirements for access for supplier's purchaser to sub-contractor premises, etc.
4.6.4 Verification of purchased documents	Expand to include surveillance of sub-contractor by supplier.
4.9 Process control	Expand to include preventive maintenance and qualify "suitable working environment" to include safety aspects.
4.13 Control of non-conforming	Expand to include system nonconformities.
4.16 Quality records	Expand to include quality audit records and verification records.

Finally, other changes call for more "specific references" or "clarification." Examples would include, clauses 4.1.1, *Quality Policy*, and 4.1.2.3, *Management Review*.

ISO and the Year 2000

Reviewing the amendments and calling for expansion, one notices that in some cases, the "upgrades" are little more than a rephrasing of text already included in the ISO 9004 guidelines. This apparent tendency to "transfer" ISO 9004 clauses into 9001/2/3 is even more evident when one reviews the 1996 proposed draft revisions of the U.S. Tag team (not to be confused with the ISO TC/176 committee). Perusing the draft one comes across the following "new" clauses:

4.1.3.3 Business Process Planning
4.2 Work and People Management Process
4.2.1.2 Work Definition
4.2.1.3 Work Planning
4.2.2 People Management
4.2.2.1 Performance
4.2.2.2 Development
4.2.2.3 Support
4.2.2.3.1 Safety
4.2.2.3.2 Security
4.2.2.3.3 Welfare
4.3 Operational Processes
4.3.1 Customer Process
4.3.1.1 Marketing
4.3.1.2 Sales
4.3.1.3 Tender and Contract Review
4.3.1.4 Customer Satisfaction
4.3.3.4.2 Customer Verification of Sub-contracted Products
4.3.7 Installation Process (three sub-paragraphs)
4.3.8 Maintenance Process (four sub-paragraphs)
4.3.9 Measurement Process (much more detailed than current version)
4.4 Operational Support Processes (four sub-paragraphs)
4.4.2 Accounting Process
4.4.3 Quality Support Techniques
4.4.3.2 Statistical Techniques (not new but specific reference to statistical process control and process capability studies)
4.4.3.3 Quality Related Economics
5.0 Audit Process (internal and external plus audit analysis)

Obviously many people have been very busy thinking up ways on how to "improve?" the ISO 9000 series. Those familiar with the Malcolm

Baldrige Award will quickly recognize that some members of the U.S. TAG Committee would like to see the ISO 9000 series adopt a "Baldrigean" philosophy. It must be realized that the above listing is but a *draft document proposed by the U.S. Tag team.* The recommendations are certainly noble and well-intentioned but will they be accepted by the international community; a community which, for the most part, already has great difficulty implementing the 1987 version of the ISO 9000 series? I do not believe so, at least not before the first decade of the next century.

15 Conclusions

Everything looks impossible to those who never try anything.
Jean-Louis Etienne

How to Get There?

One of the most common mistakes made by companies who have embarked on the road to ISO 9000 registration is to treat each ISO paragraph as independent units. This is perhaps due to the fact that when preparing the responsibility matrix, paragraphs are assigned to each division(s) without emphasizing that the certification process is a team effort. The problem with such an approach is that each division looks at ISO from the narrow perspective or point of view of "its relevant paragraph." Throughout the book I have emphasized that the best way to achieve certification is to recognize that the ISO standards offer an exceptional opportunity to work in teams of internal customers and suppliers to help achieve total quality at all levels.

Implementing an ISO 9001, 9002 or 9003 type quality system will necessitate bringing about changes. Since most people and organizations tend to resist change, bringing about changes invariably leads to stressful situations. In order to bring about successful, productive changes within an organization, everyone must find the time to modify their behavior in such a way as to accentuate positive factors and minimize negative factors. To do so, managers must adapt new models of management and avoid the classic mistakes tabulated in Table 15.1.

In the words of Rosabeth Kanter Moss, managers *must become change masters adept at the art of anticipating the need for, and of leading, productive change.* Change masters must be able to communicate and demonstrate the need for change. This can be achieved by first explaining to *all* concerned *why* change is necessary. Next, managers should take an active role in the implementation of the ISO quality system. This is usually accomplished by participating in ISO meetings, or requesting minutes from such meetings. Finally, in order to erode the well-known psychological resistances to change, management must help bring down departmental barriers. People must be convinced that the ISO certification is not yet another program which will go away "if only we can ignore it long enough." The more people participate in the certification efforts the easier the task and the greater the likelihood of passing the first audit. Some of the required steps needed to facilitate success are shown in Table 15.2.

Table 15.1 Frequent Mistakes Made During the Implementation of a Quality System

TYPE OF ERROR	CHARACTERISTICS	EXAMPLES
Superficiality	No profound changes.	Speeches, memos, videos.
Quick fix	Quality program rapidly put in place to solve a crisis.	Rapid drop in revenues leading to the intervention of a consultant for "quick fix".
Alibi	Start a quality program then move to something else. Or, "You don't understand our industry."	Hiring a quality director with no support staff! Failures attributed to the special nature of the business.
Poor Adaptation	Start-up of a quality program to solve problems of another nature.	Social unrest, low motivation, why not start a quality program?
Individualism	Everyone manages "his" quality but NOT the overall quality.	No coordination between teams. Individualized indices.
Traditionalism	"We have always done it this way, so why change?"	Application of the same old methods.
Slogans	Setting posters everywhere.	T-shirts, caps, generic memos.
Fashionable	Everyone is doing it, so why not us?	Follow the Japanese example or a competitor's successful application.
The Gimmick	Belief in the all purpose quality tools offered by certain consultants.	Quality circles, SPC, DOE, etc.
Spreading out	No global approach to quality.	A little bit of quality here and there.

Table 15.2 Organizational Strategy

Action (across) Actors (down)	Global Strategy	Daily Transformation
Upper Management	• Select Quality System(s) • Formulate long term quality policy. • Define investment and other resource commitment. • Nominate guardian(s) of quality policy.	• Act as "Change Masters". • Practice what you preach. • Become leaders/teachers in all aspects of problem solving methodology. • Innovate. • Find ways to energize the "grass roots."
Middle Management	• Internal supplier and customer of the quality policy.	• Learn how to delegate responsibilities. • Receive training and become teachers.
Operators	• Interviewed as an internal client. • Will need constant clarification of stated objectives and methodology.	• Increased responsibilities. • Improvement in the quality of work via timely training.
Customers	• Integrated within the quality policy. • Interviewed to assess his needs and requirements.	• Improved communication between engineers and marketing.
Suppliers	• Also integrated within the quality policy. • Interviewed to determine how to improve partnership.	• Better understanding of requirements/specifications. • Integrated within the process.
Experts	• Assist on overall strategy, methodology and implementation.	• Provide technical expertise. • Guidance.

Perhaps the best advice one could offer is to follow the ancient Chinese proverb which states: "Tell me and I will forget. Teach me and I will remember. Involve me and I will learn."

ISO's Recognition

As more and more national standards — over 92 worldwide — begin to recognize the ISO 9000 series (China adopted the ISO series in October, 1991), it will become increasingly difficult to deny ISO's stature as a world standard. In the U.S., several federal agencies have recognized, or are about to recognize, the ISO 9000 series. Besides the Department of Defense, the Federal Aviation Administration (FAA) and the Federal Drug Administration (FDA) are currently "studying" the implication of adopting ISO 9000. No doubt the Aerospace Industries Association and the American Institute of Aeronautics will soon be affected by ISO.

The American Association for Laboratory Accreditation (A2LA) signed a memorandum of understanding with the U.S. Environmental Protection Agency's (EPA) Environmental Monitoring Systems Laboratory which requires that its approved suppliers be registered to at least ISO 9002.

The American Gas Association (AGA) Laboratories and DQS, Germany's Quality System Registrar, mutually recognize ISO 900 registration of gas-fired product manufacturers. Finally, one should note that Japan's Industrial Standard Marking System has incorporated the ISO 9000 series into its system.

In Europe, the race towards ISO registration is in fact just beginning. Whereas it is true that the U.K. has over 30,000 firms registered to one of the three ISO standards, it is the exception rather than the rule. Indeed, with the possible exception of the Netherlands which might have as many as 12,000 to 15,000 firms registered to ISO, all other European countries are just beginning to see companies register to one of the ISO standards. In June 1990, a representative from the Association Française Assurance Qualité or AFAQ, France's nearest equivalent to the U.K.'s NACCB, proudly wrote in *Qualitique*, that France had just registered its one hundredth company and would have about 600 registered by the end of 1991. But what about Germany, Spain, Italy and the other member states? How many firms are registered to ISO in these and other European countries? Probably not more than a few hundred. Yet, whereas it is true that most European countries do not seem to be rushing towards ISO registration, recent events (December 9-11, 1991) at Maastricht, the Netherlands, would indicate that most members of the EC community are indeed focusing their efforts towards the eventual formation of a single unified economic market. Non-EEC members, including the

U.S., should take note and remember that as outsiders looking in, a different set of rules might apply when it comes to achieving ISO registration.

ISO and the EEC Directives

As the *Product Liability* and the *Product Safety Directives* issued by the European Community in July 1985 and July 1989 respectively are being transformed into national laws by each member of the member states, more and more industries will be affected. Indeed, as Walter H. Boehling explains in the June, 1990 issue of *Quality Progress*, that after a manufacturer has demonstrated (via third-party certification), that his product does "conform to essential requirements... the principle of mutual recognition will mean that, when there is an EC directive, conformance to any standard in any member state *gives the right of access to the entire free market*" [*Quality Progress* Boehling, p. 30, emphasis added].

At present, the following industries are covered by the EC directives:

- Building products
- Gas appliances
- Lifting and loading equipment
- Machine safety
- Medical devices
- Mobile machines
- Personal protection equipment
- Radio interference
- Rollover protection
- Simple pressure vessels
- Small industrial trucks
- Telecommunication terminal equipment
- Toys

If your industry belongs to one of the above groups it is very likely that *you will sooner or later* be requested by someone to achieve registration to one of the ISO 9000 series of standards. *In addition*, your product will have to conform to the applicable directive. For member states (i.e. members of the EEC), this will mean that the manufacturer's product(s) will have to bare the "CE" mark affixed by the manufacturer or an approved third-party (notified body). This mark will indicate that the manufacturer not only has a quality assurance system which has been registered with an official third-party but also that the product conforms to the requirements set by the directive. For non-member states, such as the United States, the CE mark can be obtained as long as the manufacturer submits a technical dossier which demonstrates the product's conformance to the particular directive. Some U.S. manufacturers will be able to rely on their European subsidiaries to affix the EC mark as long as the manufacturing plant is ISO registered to the appropriate 9000 quality assurance system (a 1990 survey of appliance manufacturers revealed that most U.S. (appliance) manufacturers have either design, manufacturing or distribution facilities in one or more European country, see *Appliance Manufacturer*, May 1990, pp. 64).

The issues relating to the CE mark are still being formulated in Brussels. Contrary to what is generally believed in the U.S., this is not a protectionism ploy devised by the Europeans to keep U.S. products out of the EEC. Rather, it is a means to normalize and facilitate trade within the European community (for a positive statement regarding "Europe 1992," see "Europe 1992 a Plus for Some Appliance Makers," in *Appliance Manufacturer*, May 1990, pp 64-66). European manufacturers are themselves confused and ill informed as to the nature of these EC directives. A French manufacturer of medical devices (a subsidiary of a U.S. multinational), whom I visited in September 1991, had great difficulties interpreting the legal jargon which permeates the July 1991 EC directives for medical equipment (see paragraph below). The reassuring fact is that the EC directives will have to conform to the General Agreement on Tariffs and Trade

(GATT) agreement on Technical Barriers to Trade which guarantees non-EC product the same access as EC products.

A Brief Look at Directives Concerning Medical Devices

The following is but a brief overview of a draft proposal for a Council Directive concerning medical devices. The author was fortunate to obtain the July 1991 draft document published by the *Commission of the European Communities* during a consulting assignment in France. The 89 page document is divided into twenty-four articles and twelve annexes. The document's constant cross referencing to other articles makes it difficult to read. Before briefly reviewing the draft directives, one must *emphasize* Article 12 which states that "This Directive is addressed to the Member States" [Commission of the European Communities, p. 40]. Nonetheless, even though the directive is addressed to member states, non-member producers of medical devices should perhaps be aware of its contents.

The directive subdivides medical devices into four classes I, IIa, IIb and III. The classification criteria are specified in a seven page annex (Annex 9). The rules are much too complex to include here but basically are structured in such a way as to distinguish between non-invasive, invasive, implantable, therapeutical, active and other types of devices. Rule 1 sets the tone of Annex 9, "All non-invasive devices are in Class I, unless one of the rules set out hereinafter applies (p. 76)." [The reader will note that there are thirteen other rules!] The class of product in turn determines the Conformity Assessment procedures for the CE mark. A sample taken from Article 10 will help demonstrate how tedious the directive can be.

> 1. In the case of devices in Class III other than devices which are custom-made or intended for clinical investigations, the manufacturer shall, in order to affix the CE mark, either:

(a) follow the procedure relating to the EC declaration of conformity set out in Annex 2 (full quality assurance), or

(b) follow the procedure relating to EC type-examination set out in Annex 3, coupled with:

I. the procedure relating to EC verification set out in Annex 4, or

II. the procedure relating to the EC declaration of conformity set out in Annex 5 (production quality assurance) [Draft, p. 25].

Remember, this is only for Class III devices; Article 10 continues for three more pages. No wonder my French friends were having some difficulties interpreting the document!

One must conclude by noting that a manufacturer will be able to affix the CE mark if and only if he has some type of quality assurance system in place (the directive differentiates between "Full Quality Assurance System (Annex 2)," "Production Quality Assurance (Annex 5)," and "Product Quality Assurance (Annex 6)"). Finally, irrespective of the type of quality system, "The manufacturer shall lodge an application for assessment of his quality system with a notified body." To complicate matters further, an official list of European notified bodies (third party registrars) is not yet available.

Some Pitfalls, Criticisms and Suggestions Regarding the Standards

One of the major faults that have been attributed to the ISO 9000 series is that because of its generic phraseology, uninitiated auditors can often misinterpret or worse, impose their own interpretation onto auditees. Allan J. Sayle raises some legitimate and not so legitimate concerns.

Regarding the implementation of corrective actions, Sayle correctly points out that the standard does not specify any time frame. "Accordingly, unless specified by the prime customer, a supplier can take all the time in the world to implement corrective action and still satisfy the standard." [Sayle, *Management Audits*, p. 6-3]. The fact that no deadlines for corrective actions are set by ISO 9001 should not necessarily be perceived as a weakness in the standard. Each organization should really be allowed to decide what time frame to select. Besides, some costly corrective actions may only be implemented over the span of several months or a year.

Sayle also points out that since no frequency is specified for internal audits, "An auditee needs only to perform such audits once every century to comply with the letter of the standard." Perhaps, however paragraph 5.4.1 of ANSI/ASQC Q94 (ISO 9004) does suggest that the quality system should be "audited and evaluated on a regular basis." Certainly, once every century would satisfy the requirement of a regular basis, but is it a reasonable interval to "verify whether quality activities comply with planned arrangements and to determine the effectiveness of the quality system"? [ANSI/ASQC Q91, paragraph 4.17].

Perhaps Sayle's most pertinent recommendation relates to his advice regarding interpretation of the standard.

> Over the years, there have been fruitless arguments, pointless squabbles and enmities caused by ignorant auditors attempting to interpret the meaning of such documents unilaterally and to dictate the actions which an auditee ought to take. Whenever there is any doubt, the sensible course is for auditor and auditee to agree upon a compromise understanding. This should also be the case for statutory auditors who might have the right to invoke a particular interpretation but do not have the right to abuse their authority and should always recognize that an honest

mistake can be made by an auditee [Sayle, *Management Audits*, p. 6-5].

I wholeheartedly agree.

ISO and the Service Industries

In the U.K., where over 30,000 firms have been certified to one of the ISO standards (mostly 9001 or 9002, i.e. BS 5750 Part 1 or 2), one finds that ISO has infiltrated the service industries. Bakeries, hotels, legal services as well as one man operations have been ISO certified. In 1990, the International Standard Organization released a draft document ISO/DIS 9004-2 entitled "Quality management and quality system elements — Part 2: Guidelines for services." The standard (guidelines) "is based on the generic principles of ISO 9004 and provides a comprehensive overview of a quality system specifically for services" [ISO/DIS 9004-2, p. 3].

Although the standard "can be applied in the context of developing a quality system for a newly offered or modified service," I would question whether the quality system really "embraces all the processes needed to provide an effective service from marketing to delivery and the analysis of service provided to customers" [ISO/DIS 9004-2, p. 3]. I do not so much question whether or not the standard does or does not provide all that is required "to provide an effective service," for indeed it does seem to satisfy that requirement, but rather whether one can view the service industry as a monolith. Surely, there are differences between a service industry providing its customers with hamburgers, and the one providing open heart surgery. Besides the obvious different level of expertise and training on the part of the provider of the service, the customer cannot request the same set of expectations simply because he/she would have to trust his pallet in one case and a doctor's opinion in the other. One can also ask if a restaurant can be compared to an aluminum plant? To some extent yes, but the restaurant in essence produces its own specifications which the customer must then match to his/her liking. If the match is favorable, the customer returns, if not, he looks for another restaurant. In some respect, the service industry has a much tougher job satisfying its customers since — unlike the aluminum plant — its customer base is broader and hence more varied.

The issues are beyond the scope of this book. It is difficult to predict if the service industry in this country will embrace the ISO 9000 series of standards with the same fervor as its British counterpart. To do so could only benefit the industry in general. Hospitals, hotels, airlines, insurance companies, gas stations etc., could certainly learn a thing or two by reading ISO/DIS 9004-2. Only time will tell [for additional reading on the service industry see articles by Carol King, Raghu Kacker and Jacques Horovitz and Chan Cudennec-Poon listed in the Bibliography].

Consulting and ISO 9000: A Few Words of Caution

When, in the summer of 1989, I first mentioned the ISO 9000 series to several of the consulting firms with whom I was sub-contracting, not one of them expressed an interest in what I had to say. But when I returned in late 1990 from a seven month consulting assignment in France, I was amazed at the exponential growth of U.S. experts on ISO 9000. The ISO 9000 series had finally landed in the U.S., at least in some cities. Although there still remain vast regions of near total indifference towards ISO (particularly in western and mid-western states), ISO 9000 awareness is rapidly spreading westward and southward across the U.S.

Anyone who subscribes to one or more of the quality periodicals will no doubt have noticed that every month a new consulting group is offering its ISO 9000 expertise. Faced with the increasingly vast array of expert opinions and seminars on ISO, how can the customer make a wise decision should the time come to select an "expert" or a seminar? I will attempt to answer that question.

The current organizational structure ("maze?"), facing each potential customer/user of the ISO 9000 series can be depicted as follows:

User = Customer
Consultant/Consulting Agency
Registrars Interpreters of the Standards
ISO (9001,2,3) Standards

There are essentially four "players." The user/customer, the consultant/consulting offices, the registrars and the ISO series of standards itself. Until now, this book has focused on three of the four players: customer, registrars and the standard itself. What about the consultant/consulting offices, how do they fit within the structure?

Types of Consulting and Services

There are basically eight types of consulting services, they are:

- Management or accounting firms who have recently added ISO 9000 consultancy to their list of "expertise." This is usually achieved (almost overnight) by simply hiring an "expert" or acquiring a consulting firm which *might* already have the relevant expertise.

- Consulting firms which have seen the need to upgrade and expand their practice.

• Universities, state universities or colleges who also wish to join in (very competitively, I might add) the ISO 9000 market. This is usually achieved via the universities extended education program.

• Individual consultants with varying degrees of expertise who either directly contract their skills or sub-contract to one of the above institutions.

• Product inspection/testing organizations who have upgraded their services by including a division in charge of quality system inspection. These organizations, of which there are very few, usually advertise their services as facilitating the ISO 9000 implementation process (i.e. they will write and organize your quality system to suit your needs!).

• Organizations which specialize in ISO 9000 lead auditor or similar courses. These organization are invariably headquartered in the U.K. or a commonwealth member nation.

• Third party registrars who, after much soul searching, have finally decided to join the ranks of consultants. Some registrars achieve this delicate *tour de force* by creating a *separate* consulting branch within the organization.

• Finally, one is left with consulting firms whose members usually are ex-ISO 9000 auditors/assessors. Most of these firms, which tend to be headquartered overseas, offer seminars on lead auditor training and internal auditors training.

All of the above agencies will generally offer all or some of the following services or combination of services:

- In-house or public seminars.
- Five day lead auditor training courses.
- Internal auditing courses.
- Pre-assessement audit.

- Third party audits (only registrars).
- Manual writing (quality or any other manual).
- Executive seminars.
- Other ingenious combinations of the above.

Faced with the plethora of options, what should you do? Although it is very difficult to offer a recommendation, I can offer some advice on what you should try to avoid doing. Before selecting a consultant or a consulting organization you should first ask a few questions. At a minimum these questions should help you:

- Assess the consultant's expertise (weeks, months or years).
- Determine the number of companies the consultant/agency has assisted.
- Check on a couple of references.

What Services Should You Contract For?

It is my conviction that companies should only sub-contract for certain types of ISO 9000 consulting services. There are basically two things that a company should *never* ask a consultant to do:

- Ask him/her to write a quality manual.
- Ask him/her to write the quality assurance system.

The reasons should be obvious. If a consultant is hired to put a system in place, then the consultant better be present on the days of the third party audit for he will likely have become the one expert with the most knowledge about "your" quality assurance system. In such cases, your chances of passing a third party audit are actually significantly reduced. Another disadvantage of hiring a consultant to document your quality assurance system is that it usually is somewhat expensive. Naturally this is good for the consultant. When you hire a consultant to document your processes, including your quality manual, you must pay him/her to first understand the "ins and outs" of your industry. This may easily take as much as five to seven days, assuming you have an experienced consultant who can quickly grasp the subtleties and complexities of your particular industry. Besides, why would you want to pay someone to tell him/her what and how you do something, so that he/she can then write it down for you? You in turn will have to edit what has been written down to ensure that it is all correct! Why indeed would you want to pay for all that "service" simply because you don't have time to do it? If indeed your management cannot allocate enough time or resources for you to ensure that your ISO implementation process will succeed, then why would it be willing to spend tens of thousands of dollars in consulting fees?

Naturally, as a consultant, I could not conclude this section without offering some positive comments regarding the role of a consultant. Based on my experience, a consultant can be most effective in a variety of roles. He/she can:

- Offer training
- Offer recommendations based on his varied experience.
- Act as a catalyst to motivate and direct people.
- "Float" across departments. This activity can be particularly helpful in situations where inter-departmental rivalries are strong.
- Act as an interface (similar to the above point).
- Guide and help monitor efforts via carefully planned periodic visits. This activity ensures that the overall implementation plan is kept "under control."
- Answer difficult questions.
- Review and edit but NOT write documentation (manuals, etc.).
- Conduct pre-assessment audits and offer unbiased constructive opinions/suggestions.

The above activities should not require an inordinate number of scheduled days. Certainly, the amount of consulting activities will depend on the size of the facility. However, since the emphasis is placed on the consultant guiding the client and offering recommendations on how the quality system should be organized, documented and verified, most consulting jobs should last anywhere between eight and twenty days broken down as follows:

- Two to six days training (i.e., presenting a detailed seminar on the ISO 9000 series (what it is?, why it is required?, what is entailed?, how to begin?, etc.) to selected members of the company's staff.

- Two to four days recommending how to organize the system or monitoring the efforts.

- The balance should be reserved for internal audit assessment as well as internal auditing training, and monitoring of corrective actions. The assessment activity could be split into pre- (three to five days) and post- (three to five days) auditing.

Naturally, the above recommendations are but one man's opinion.

Finally I would like to conclude by quoting from the excellent book by Michael L. Dertouzos, Richard K. Lester and Robert M. Solow entitled: *Made in America: Regaining the Productive Edge.*

> To understand what has been happening to American productivity, one must know what has been happening on the shop floor, in the laboratory, in the boardroom, and in the classroom (p. 3)... The international business environment has changed irrevocably, and the United States must adapt its practices to this new world (p. 8)... Specifically, we think that failures by American firms and industries to adapt to new conditions have played an important role. Some of these shortcomings are deeply rooted in organizational structures and social attitudes, and

> they will be at least difficult to put right as any macroeconomic problems (p. 38). [For a more optimistic opinion as to the future of the U.S. economy, see John Naisbitt and Patricia Aburdene's *Megatrends 2000: Ten New Directions for the 1990's*.]

Having had the opportunity to consult with American and European multinationals and live in Europe for two years (1989-1990), I would readily agree with the authors' assessment summarized in the above quotation. I would further add that, contrary to popular belief, the next industrial challenge is not likely to come from Japan but may instead come from Western Europe (see for example, Daniel Burstein's *Euroquake* published by Simon & Schuster). I would therefore suggest that if you do indeed perceive the ISO 9000 registration effort as an opportunity to bring about long needed improvements, the investment will be well worth the many rewards. Teamwork and registration to either 9001, 9002 or 9003 will most likely facilitate your next venture, be it a national award or some other award. If you have read that far along, my parting advice would be: *don't delay, start today*.

Bibliography

Dennis R. Arter. *Quality Audits for Improved Performance*, 1989.

Daniel Burstein. *Euroquake*. Simon & Schuster, 1991.

W.H. Boehling, "Europe 1992: Its Effect on International Standards," *Quality Progress*, June 1990, pp. 29-32.

Michael L. Dertouzos, Richard K. Lester and Robert M. Solow. *Made in America*. Harper Perennial, MIT Press, 1990.

John P. Graham, "Texas Petrochemical plant gets certification," *The Oil & Gas Journal*, May 13, 1991.

H. James Harrington. *Poor-Quality Cost*. ASQC publication (1987).

Jacques Horovitz and Chan Cudennec-Poon. "Putting Service Quality into Gear," *Quality Progress*, January 1991, pp. 54-58.

J. Ladd Greeno, Gilbert S. Hedstrom and Maryanne DiBerto. *Environmental Auditing: Fundamentals and Techniques, 2nd edition*. Center for Environmental Assurance, Arthur D. Little, Inc.. 15 Acorn Park, Cambridge, MA 02140. Phone (617) 864-5770 X-2433.

International Standard ISO 10011-1 1990-12-15. Guidelines for auditing quality systems — Part 1: Auditing.

ISO 10011-2 1991-05-01. Guidelines for auditing quality systems — Part 2: Qualification criteria for quality systems auditors.

ISO 10011-3 1991-05-01. Guidelines for auditing quality systems — Part 3: Management of audit programmes.

J.M. Juran and Frank M. Gryna. *Quality Planning and Analysis* (2nd edition). McGraw-Hill Book Company, 1980.

Raghu N. Kacker. "Quality Planning for the Service Industries," *Quality Progress*, August 1988, pp. 39-42.

Carol A. King. "A Framework for a Service Quality Assurance System," *Quality Progress*, September 1987, pp. 27-32.

James Lamprecht, "Demystifying the ISO 9000 Series Standards." To be published in *Quality Engineering*, Volume IV, no. 2, 1992.

James Lamprecht, "The ISO Certification Process: Some Important Issues to Consider," in *Quality Digest*, August, 1991.

James Lamprecht, "ISO 9000: Strategies for the 1990s," in *Quality Magazine*, November, 1991.

Donald Marquardt et al., "Vision 2000: The Strategy for the ISO 9000 Series Standards in the '90's," *Quality Progress*, May 1991, pp.25-31.

Charles A. Mills. *The Quality Audit: A Management Evaluation Tool*, 1989.

David J. Morrow, "Be Good Or," *International Business*, July 1991, pp. 24-28.

John Nisbett and Patricia Aburdene. *Megatrends 2000. Ten New Directions For the 1990's.* Avon Books, 1990.

Michael E. Porter. *The Competitive Advantage of Nations.* Free Press, 1990.

Norman C. Remich, Jr, "Europe 1992 a Plus for Some Appliance Makers," *Appliance Manufacturer*, May 1990, pp. 64-66.

Charles B. Robinson. *How to Plan an Audit*, 1987.

Allan J. Sayle. *Management Audits: The Assessment of Quality Management Systems, 2nd edition*, 1988.

Gene Schindler and James Lamprecht, "The Significance of the International Standards Organization ISO 9000 Series for the Drilling and Production Related Businesses," *The Oil & Gas Journal*, May 6, 1991

David Sloan and Scott Weiss. *Supplier Improvement Process Handbook*, 1987.

Dorsey J. Talley. *Management Audits for Excellence*, 1988.

Walter Willborn. *Audit Standards: A Comparative Analysis*, 1987.

Walter Willborn. *Quality Management System: A Planning and Auditing Guide*, 1989.

Malcolm Baldrige information can be obtained from:

United States Department of Commerce
National Institute of Standards and Technology
Route 270 and Quince Orchard Road
Administrative Building, Room A537
Gaithersburg, MD 20899

Members of the chemical and related industries might be interested in joining QUALICHEM, founded by the Belgian Center for Total Quality Management for the Chemical and Related Industries: QUALICHEM, pla Bedrijvencentrum Zaventem, Leuvensesteenweg, 613, 1930 Zaventem Zuid 7, Belgium. Ph: 02-759-9274.

Appendix A: Guidelines on What to Include in a Quality Manual

1.0 Mission or Policy Statement from the CEO

This statement must indicate a decision and commitment from the highest level of management. As such it should:

state the quality policy of the organization.
explain the objective of the quality manual.
confirm the C.E.O's engagement toward the stated quality policy.
ensure that the procedures stated in the quality manual are enforced.
facilitate the development of a management committed to quality.

The CEO can delegate some of the responsibilities relating to the implementation of procedures as described in the quality manual (usually a Quality Director).

1.1 Update of the Quality Manual

The update sheet should include:

the date of the first edition.
revision dates.
page numbers that have been modified and cross-references to previous editions.

1.2 Management of the Quality Manual

A brief explanation as to the steps taken by the organization to prepare, update, distribute and archive the manual should be included.

1.3 Purpose and Field of Application of the Manual

The quality manual should present an overview of the organization's general procedures designed to achieve product quality.

1.4 Definitions and Terminologies

All abbreviations should be defined.

1.5 Quality Policy and Generalized Quality Objectives of the Company

1.6 Presentation and Organization of the Company

This section could include:

Generalized financial information.
Site of headquarters.
Geographical locations of other sites and subsidiaries, etc.
Organizational chart (hierarchical and functional):

> For each division listed in the organizational chart, a brief explanation as to its function and responsibilities should be included. *[Note: Some companies prefer to include these sections in the second tier document.]*

1.7 Responsibilities with Regard to Quality

This section should contain a list of all individuals responsible for quality by department, process and/or activities.

2.0 Product Design and Customer Specifications

Customers needs can be identified either through (1) marketing studies or (2) direct customer orders. In either case the resulting product must:

(a) satisfy customer requirements and,
(b) satisfy precise (manufacturing) specifications, delivery dates and a host of other requirements.

To address these issues, sections of the quality manual should explain/outline who does what and how. Answering the following set of questions should helping writing each sub-paragraph.

2.1 Marketing Studies

Who (or which department) identifies the need for new products?

Surveys, data analyses.

Performance standards, life cycles, functional definitions.

Who (which department) identifies product specifications?

Ergonomics, quality characteristics.

Who (which department) listens to the Voice of the Customer?

Who is responsible for the verification of all of the above specs?

2.2 Bids/Contract Review

Where are they received?

Who verifies their content for accuracy and completion?

Who (and how) defines, verifies and negotiates:

Contractual agreements.

Technical requirements.

Issues relating to quality and quality assurance?

Others?

Upon acceptance, who receives copies of the final contract agreement?

Who is responsible for all of the above?

2.3 Design and Development Planning

Who (which department) is responsible for research?
How are investment strategies planned?

2.4 Quality Plans

When is a quality plan established?
For what type of products are quality plans developed?
Who decides when a quality plan is needed?
Who is responsible for:
The maintenance of the quality plan:
updates,
archiving,
process description,
etc.

2.5 Design Control

Design invariably involves development planning, prototypes or samples manufacturing, organizational and technical interfaces, activity assignment, design input and outputs, design verification and changes. In addition, documentation control such as approval, change modifications, etc., is also required.

How are purchasing, marketing, manufacturing, quality control and prevention, and servicing linked to obtain quality of design?

How is the necessary information provided?

How are technical norms satisfied?

Are FMEA, reliability and/or value analyses performed and by whom?

Are CAD/CAM systems used?

Who reviews/verifies (input/output) designs?

Who approves design changes?

How are these changes identified?

2.6 Document Control

How are documents maintained?
Who establishes the distribution list?
How are documents distributed?
How are outdated documents removed from circulation?

2.7 Design Changes

Who has the authority to approve design changes?

How are solutions to design changes validated?

How are design changes/revisions documented, archived?

How do you ensure that requests for changes match manufacturing changes, what control procedure is in place?

3.0 Suppliers

In order to ensure that purchased products conform to specified requirements, the organization should have procedures to:

1. Ensure that purchasing data are clearly defined and adequate.
2. Select a supplier capable of meeting requirements specified in (1.).
3. Ensure that all documents are correctly processed.

3.1 Purchasing Data

Who is responsible for establishing clear specifications as they relate to a product?

Who ensures that relevant clauses relating to overall quality and quality assurance are included?

Who specifies delivery specs and/or special conditions?

Who defines receiving inspection plans (if any)? If none, why not?

Who defines/specifies in-process inspection plans (if any)? If none, why not?

Who defines what action(s) needs to be taken in case of:

non-conformities?
modifications?
waivers?
special processes?

3.2 Assessment of Sub-contractors

Who surveys prospective suppliers?

Who evaluates suppliers and how are suppliers evaluated?

How are suppliers certified?

What records are kept on supplier performances?

Are suppliers ever removed from the certified list?

3.3 Verification of Purchased Product

Who writes purchasing orders and how are all criteria defined?

Are purchasing orders written according to well-defined quality assurance criteria?

To whom are purchasing order forms circulated?

4.0 Production-Process Control

For all other sections refer to ISO 9001, 9002, 9003 and 9004 documents.

Note: The following sample pages represent but a few pages of a quality manual. It is an amalgamation of several manuals. Most manuals should not exceed 25 to 30 pages.

Dusseldorf Inc.

Centertown Plant

Quality Manual

Table of Contents

1.0 Revisions

Date	Page	Paragraph	Comments	Approval
6/12/91	4	Paragraph 7.	Addition	John Murdock
10/21/91	2		Addition of Note	John Murdock

Approved by: John Dusseldorf	Written by: Al Lewis
Signature:	Signature:
Issued by: Quality Department	Issue Date: 11/9/91
Version: 1.3	Page 1 of 11

2.0 Distribution List

2.1 Distribution List

All distributed copies of the Dusseldorf Quality Manual are issued, controlled and distributed to the following department heads by the Quality Manager (see section 2.0).

Department	Personnel
1. Administration Bldg.	Plant Manager
2. Laboratory (Lab. 10-122)	Chief Chemist
3. Quality	Quality Manager
4. Maintenance (Bldg 11-104)	Maintenance Director
5. Human Resources (Admin. 05-133)	Human Resource Mng.
6. *Other departments as required*	

Note: Manual may be borrowed, reviewed or otherwise photocopied by anyone and thus may not always reside permanently at said location. The person in charge of said department is responsible for the manual and its subsequent circulation. Photocopies of the manual are not controlled.

Approved by: John Dusseldorf	Written by: Al Lewis
Signature:	Signature:
Issued by: Quality Department	Issue Date: 11/9/91
Version: 1.3	Page 2 of 11

3.0 Company History

In this section, a brief description can be inserted to include the plant's history. The section need not be more than a few paragraphs long. It is also a good idea to include a paragraph entitled ***Scope of Services.***

In 1957, A.G. Dusseldorf founded a business in Centertown, New Mexico. Originally located in downtown Albuquerque, the plant soon expanded into two disciplines — Building and Mechanical Engineering. By the mid 1960s, the mechanical engineering company — known as Dusseldorf Industries — was engaged in the manufacture of hydraulic pumps and cylinders primarily for use within the group but also for sale on the open market.

It was the recognition of the markets for low speed, high torque drives in the construction industry, mining and agriculture which prompted the design of a new technology hydraulic motor.

Schematics of the plant(s) can also be included, as well as the geographical distribution (if applicable) of other plants.

Approved by: John Dusseldorf	Written by: Al Lewis
Signature:	Signature:
Issued by: Quality Department	Issue Date: 11/9/91
Version: 1.3	Page 3 of 11

4.0 Manual Control

4.1 Manual Control Policy

1. Although anyone can submit revisions to the quality manual, all revisions must be approved by the plant manager.

2. The quality manager has been delegated to ensure that the latest version of the manual is distributed to all interested parties.

3. Obsolete copies of the quality manual are first retrieved by the quality manager **before** issuance of a new updated version.

4. Controlled copies of the quality manual are printed on special grey company paper.

5. All revisions are logged in the *Revision* page.

6. Versions of the quality manual are identified by a two digit version number. The first number indicates the current revision, the second the current update. Version 1.2 thus means that the manual is at revision status 1, second update. Every 10 updates a new revision is issued.

7. Holders of the manuals are responsible for assuring that the manual assigned to them is readily accessible to authorized personnel.

Approved by: John Dusseldorf	Written by: Al Lewis
Signature:	Signature:
Issued by: Quality Department	Issue Date: 11/9/91
Version: 1.3	Page 4 of 11

5.0 Management Responsibility

5.1 Quality Policy

The Dusseldorf company's quality policy (Reference QP-1) recognizes that in order to meet its customers' requirements, every employee must be allowed to participate, improve or otherwise modify processes with the intent of better satisfying customer requirements.

5.2 Responsibility and Authority

Although the plant manager is ultimately responsible for the final quality of the product, daily activities and responsibilities directly affecting a product's quality have been delegated to the divisional managers and all those reporting to them.

5.2.1 Plant Manager

A one paragraph statement describing the responsibility of each manager can be included here. The sequence could include: Quality Manager, Laboratory, Purchasing, Technical Manager, Maintenance, Human Resources, Engineering, etc.

Following the ISO/Q90 series format, the next step would be to address issues concerning Verification Resources and Personnel, Management Representative, Review, etc.

Reference: QP-1 (*Some quality manuals cross-reference other — tier two or tier three — documents. Such a format is preferred by some registrars. However the decision as to whether or not you should cross-reference is up to you.*)

Approved by: John Dusseldorf	Written by: Al Lewis
Signature:	Signature:
Issued by: Quality Department	Issue Date: 11/9/91
Version: 1.3	Page 5 of 11

6.0 Contract Review

6.1 Contract Policy

All contract reviews are performed by headquarters located in Maryville, AZ. The Centertown facility does not actively participate in these contract reviews except for the verification of technical requirements which are transmitted to the Laboratory.

Contract specifications consist of a 20 digit number which identifies several parameters including customer code, specification and other characteristics (see tier two entitled: "Work Order Specifications").

6.2 Production Planning

Production planning is organized by the engineering department which is in direct contact with headquarters via the company's computer network system. Records of all transactions are maintained in the company's mainframe computer.

Approved by: John Dusseldorf	Written by: Al Lewis
Signature:	Signature:
Issued by: Quality Department	Issue Date: 11/9/91
Version: 1.3	Page 6 of 11

7.0 Document Control

7.1 Document Approval and Issuance

Whenever practical, all documents follow the same control format as the quality manual. Exceptions would include some tier three documentation where a simplified control procedure has been adapted (see respective department for examples). All documents are issued by the appropriate department heads and are maintained by each department.

7.2 Document Changes/Modifications

All document changes and modifications are reviewed by authorized personnel (managers, supervisors and in some cases operators) in each department. Prior to issuance of new document, obsolete documents are removed by authorized personnel.

Approved by: John Dusseldorf	Written by: Al Lewis
Signature:	Signature:
Issued by: Quality Department	Issue Date: 11/9/91
Version: 1.3	Page 7 of 11

9.0 Process Control

9.1 Process Monitoring

Statistical process control forms an integral part of process control. Key variables affecting the process performance are continuously monitored via on-line data retrieval (see 18.0 Statistical Techniques).

9.2 Process Updates/Modifications

Adjustments to the process are the direct responsibility of the process team. All key processes have been documented as Standard Operating Procedures which can be retrieved at any of the many computer terminals available throughout the plant. Deviations from SOPs are handled by the supervisor and his team. All deviations are automatically stored in the mainframe data base. Quality Action Teams periodically (usually twice a month), review the deviation files and submit recommendations on how to improve the process (see Procedure QAT SOP-IMP).

Cross-reference: Corrective Action Procedures (many documents).

Approved by: John Dusseldorf	Written by: Al Lewis
Signature:	Signature:
Issued by: Quality Department	Issue Date: 11/9/91
Version: 1.3	Page 8 of 11

11.0 Inspection Measuring and Test Equipment

11.1 Types of Instruments

At least three major categories of instruments are recognized: Laboratory instruments, process instrumentation which measure key process parameters and all other equipment including safety instrumentation. Irrespective of the type of instrument, all equipment follows a preventive maintenance schedule (see PM tier two documentation).

11.2 Laboratory Instrumentation

All laboratory instruments follow a rigorous program of inspection which include: calibration, gauge repeatability and reproducibility (for some instruments), and accuracy. In some cases calibration is performed by outside State certified laboratories. Records of all instrument maintenance are stored in the laboratory's own PC network.

Reference: Laboratory tier two and tier three documents.

11.3 Process Analyzers

All instruments monitoring key process variables (see Engineering Documentation for list of parameters), are placed on a routine maintenance program for calibration and (some) for precision. Preventive maintenance is highly computerized and uses a software developed in-house by the MIS department (see MIS DOC 1-12).

References: Laboratory Documentation, MIS and PM documentation.

Approved by: John Dusseldorf	Written by: Al Lewis
Signature:	Signature:
Issued by: Quality Department	Issue Date: 11/9/91
Version: 1.3	Page 9 of 11

17.0 Internal Audits

17.1 Frequency

Random, unannounced internal audits are conducted at least once a year by trained personnel from the quality department and accounting. Each individual from the audit team is a certified ISO 9000 assessor.

17.2 Format and Records

The internal audit team follows accepted audit practices (see reference). Records of all audits, including nonconformities and corrective action(s) taken are maintained by the quality managers. A copy of all such audits is distributed to all individuals included in the quality manual distribution list.

Reference: Internal Audit Procedure.

Approved by: John Dusseldorf	Written by: Al Lewis
Signature:	Signature:
Issued by: Quality Department	Issue Date: 11/9/91
Version: 1.3	Page 11 of 11

Appendix B: Quality System Registrars (December 1991)

ABS Quality Evaluations, Inc.
Robert C. Sutton, President
263 North Belt East
Houston, Texas 77060
Ph: (713) 873-9400
Fax: (713) 873-9364

American Gas Association Laboratories (AGA)
8501 E. Pleasant Valley Rd.
Cleveland, Ohio 44131
Ph: 216-524-4990

ATT
650 Liberty Ave.
Union, NJ 07083
Ph: 800-521-3399
Fax (908) 851-3360

Bureau Veritas Quality International
Walter F. Buesing/K.C. Wal
509 North Main Street
Jamestown, N.Y. 14701
Ph: (716) 484-9002 Fax: (716) 484-9003

Canadian General Standards Board
(Office) 9C1 Phase 3
Place du Portage
11 Laurier Street
Hull, Quebec, Canada

(Mail) Qualification and Certification Listing Branch
Canadian General Standards Board
Ottawa, Canada K1A 1G6
Ph: 819-956-0439

Det Norske Veritas (DNV) Industrial Services, Inc.
Yehuda Dror
16203 Park Row, Suite 160
Houston, Texas 77084
Ph: 713-579-9003

ETL Testing Laboratories/
Industrial Park
Cortland, New York 13045
Tel: 502-ISO-9000 or 607-753-6711
Contact: Mr. J.R. Williams (ETL)

Intertek Service Corporation (ISC)
9900 Main Street, Suite 500
Fairfax, VA 22031
Tel: (800) 336-0151

Lloyd's Register Quality Assurance Ltd.
c/o Lloyd's Register Shipping
David Hadlet
17 Battery Place
New York, New York 10004
Ph: 212-425-8050

Quality Management Institute
Malcom J. Phipps
1420 Mississauga Executive Center
4 Robert Speck Parkway
Mississauga, Ontario L4Z 1S1, Canada
Ph: 416-272-3920

Quality Systems Registrars, Inc.
Cass Tillman
P.O. Box 55129
Metairie, LA 70055
Ph: 594-455-1602

SGS Yarsley Quality Assured Firms
Emani Pires
1415 Park Avenue
Hoboken, New Jersey 07039
Ph: 201-792-2400

TUV Rheinland of North America, Inc.
Joseph DeCarlo or Dr. Klaus Spiegel (President)
12 Commerce Road
Newton, CT 06470
Ph: 203-426-0888

Other TUV offices in: Austin, TX (512) 343-6231; Longwood, Fl (407) 774-1222; Livonia, MI (313) 464-8881; San Ramon, CA (415) 820-8444; Marlborough, MA (508) 460-0792; San Diego, CA (619) 792-2774 and North York, Canada (416) 733-3677.

Underwriters Laboratories, Inc.
Charles Mauro or Michael Caruso
1285 Walt Whitman Rd.
Melville, N.Y. 11747
Ph: 516-271-6200

Underwriters Laboratories, Inc.
Peter Chresanthakes/Jeff Dohrmann
333 Pfinsdten Road
Northbrook, IL 60062
Ph: 708-272-8800

Underwriters Laboratories, Inc.
Stuart Walker/Steve Schmid
12 Laboratory Dr.
Research Triangle Park
North Carolina 27709
Ph: 919-549-1400

Underwriters Laboratories, Inc.
Beatrice Lee/Bruce Santo
1655 Scott Blvd.
Santa Clara, CA 95050
Ph: 408-985-2400

Vincotte USA, Inc.
Carl M. King
10497 Town & Country Way, Suite 900
Houston, Texas 77024
Ph: 713-465-2850

Other Useful Phone Numbers and Addresses

American National Standards Institute (ANSI)
11 West 42nd Street, New York, New York 10036
Ph: (212) 642-4900 (Sales)

BSI
2 Park Street
London, W1A2BS
England
Ph: (011) (44) 71 629-9000
Contact: Mr. Keith Wicks

Note: BSI publishes the ***Association of Certification Bodies*** (*not all of which are NACCB accredited*).

Institute of Quality Assurance (IQA)
10 Grosvenor Gardens, London SW1W 0DQ
United Kingdom

Department of Trade & Industry
Room 312
Kingsgate House
68-74 Victoria Street
London SW1E 6SW
England
Ph: (011) (44) 71 215-8123

International Standards Organization (ISO) (212) 642-4900

National Accreditation Council for Certification Bodies
3 Birdcage Walk
London
SW1H9JH, England
Ph: (011) (44) 71 222-5374

Note: The NACCB publishes a ***Directory of Accredited Certification Bodies***.

National Institute of Standards and Technology (NIST)
Ph: (301)975-4038

National Technical Information Service
Springfield, VA 22161
Ph: (301) 975-3058

Registrar Accreditation Board (RAB)
310 West Wisconsin Avenue, Milwaukee, WI 53203
Ph: (414) 272-8575

Sources for Ordering Standards

American Society for Quality Control
ANSI/ASQC Q90-Q94. The complete set can be ordered for $44.95 ($40.50 for ASQC members) from:

> ASQC Customer Service Department
> P.O. Box 3066
> Milwaukee, WI 53201
> Ph: (800) 952-6587

American National Standards Institute (ANSI)
1430 Broadway
New York, New York 10018,

Foreign/International: (212) 642-4954
Domestic: (212) 624-4954
Fax: (212) 302-1286 or 398-0023

Information Provided: ANSI and ANSI approved industry Standards, international and foreign standards.

Global Engineering Documents
2805 McGraw Avenue, P.O. Box 19539
Irvine, CA 92714
Telephone: (800) 854-7179 or (714) 261-1455
Fax: (714) 261-7892

Information Provided: Industry standards, federal standards and specifications, military standards and specifications and foreign standards.

National Standards Associations (NSA)
1200 Quince Orchard Blvd.
Gaithersburg, Maryland 20878
Telephone: (800) 638-8094
(301) 590-2300
Fax: (301) 990-8378

Information provided: Industry and military standards and specifications.

General Services Administration (GSA)
Specifications Branch
Seventh and D Streets, S.W.
Washington, D.C. 20407
Telephone: (202) 708-9205
Fax: (202) 708-9862

Information specified: Federal standards and specifications.

Appendix C: How to Use the ISO Questionnaire

The questionnaire was designed by the author to help him assert a company's readiness to achieve certification. The questionnaire is most effective if the following conditions are satisfied:

- The auditor/assessor knows how to conduct a third party audit. This can be achieved by attending and passing the lead assessor exam offered by one of the many (approved) five day seminars advertised in *Quality Progress*.

- Passing the lead assessor exam does not automatically qualify you as a (professional) auditor. You must also have conducted at least five audits to fully understand how the auditing process works (certainly not required for internal audits).

Naturally, not everyone will have satisfied these requirements prior to using the ISO questionnaire. This should certainly not prevent you from using the questionnaire for your pre-assessment. Before using the questionnaire, you should however ensure that every "auditor" has at least read Chapters 12 and 13.

How to Administer the Questionnaire

Elect a lead auditor. The best candidate would probably be the quality director or manager. The lead auditor should in turn nominate two to three individuals to act as "auditors" to help him perform the audit.

How to proceed?

- Plan your "audit-trail." Decide how you will proceed, and who shall ask what questions. Since this is likely to be your first ISO audit, I would urge you to proceed as a team. The team approach will allow team members to support each other.

• Explain the purpose of the audit and be sure to stress that this is a fact finding audit designed to help you (the team) decide how much effort will be required to implement an ISO quality system.

• Try to avoid reading the questionnaire. This will be difficult at first since you do not know its content. However as one member of the team starts asking questions the other members can read ahead and begin formulating their next series of questions. Remember that the questionnaire should really be perceived as nothing more than a script which helps you conduct your audit. If you simply read the questionnaire, the audit can quickly take on the appearance of an inquisition (something you definitely want to avoid).

• As you begin collecting information, each team member will rate the questions (see introduction to questionnaire for an explanation of the grading scale). After every couple of sessions, take a break, find a quiet room and compare notes. There are basically two procedures available. You can either:

(a) Survey each team member for each question and discuss the individual ratings. Reach a consensus and grade. Repeat for each question.

Advantages: This technique allows each member to understand how the other members interpret ISO. The discussions are very valuable in that by the end of the questionnaire, team members will have a better overall understanding of the process and the grading process will become more standardized. As with any grading procedure, you will find that some of your team members are tough graders whereas other tend to be lenient.

Practice your team problem solving skills and ensure that everyone is listened to.

Disadvantages: The process is slow and may well require as much as thirty minutes or more per session.

(b) Survey each team member for his/her score. Compute the average and standard deviation or range. The smaller the range, the better the consensus and vice-versa. Discuss questions with ranges greater than or equal to two points.

Advantages: Quick and easy to use. Fans of Statistical Process Control can even chart the voting process. The sample size (n) would be 4 (i.e. four auditors), and the number of samples could be the total number of questions or an average of the questions making up each section. The target score should probably be set at 4.5 with a three standard deviation spread of 0.5 (on either side of the mean). Scores below 4.0 would indicate areas needing attention.

Disadvantages: No or little discussions.

- When you are finished, use the circular graphics to summarize results form your audit (see example).

ISO 9001-2 Quality System Questionnaire

CRITERIA: Please use the scale described below to evaluate each question. If you don't know or the item is not applicable, simply write DK or N/A. ISO 9002 firms should ignore sections 5.0 Design Control and 19.0 Servicing.

(0) This element is not included in the supplier's quality system.

(1) This element is outlined in the supplier's quality system, but both planning and execution require substantial improvement.

(2) The element is included in the supplier's quality system and planning/procedures are generally adequate. However, execution requires improvement.

(3) This element is included in the quality system and is generally well-executed. However the level of planning is inadequate and requires further improvements.

(4) The element is included and execution and planning are adequate.

(5) The element is included and exceed requirements. The supplier has innovated in this area.

Note: This questionnaire is only meant to be used as a guideline.

Ratin

1.0 MANAGEMENT RESPONSIBILITY

1.0 With regard to quality, have you defined your policy, objectives and commitment? ____

2.0 Have you documented your policy, objectives and commitment to quality? ____

3.0 Is your quality policy understood, implemented and maintained at all levels in the organization? ____

ORGANIZATION

4.0 Have you defined the responsibilities and authority of:

4a. person(s) involved in the final testing, control and verification requirements? ____

4b. person(s) involved with the identification and recording of any product quality problems? ____

5. Have the roles, responsibilities and authority of person(s) involved with the following activities been defined: ____

Rating

5a. internal audits. ______

5b. design reviews. ______

5c. monitoring of the design process. ______

6. Are the design reviews, audits (quality system, procedures, product) and other inspection procedures carried on by persons other than those directly involved with the aforementioned tasks? ______

7. Has a management representative been assigned to ensure that the ISO 9000 requirements are implemented and maintained? ______

8. Is the quality system periodically reviewed at appropriate intervals to ensure its continuing suitability and effectiveness? ______

9. Are records of the aforementioned reviews maintained? ______

2.0 QUALITY SYSTEM

10. Do you have a quality manual? ______

11. Have you established written procedures in the following areas:

11a. Preparation of quality plans in accordance with specific customer requirements. ______

11b. Updating, as necessary, of quality control, inspection, and testing techniques, including the development of new instrumentation. ______

11c. Verification of design compatibility. ______

11d. Identification and acquisition of any controls, processes, inspection equipment, fixtures, total production resources, and skills that may be needed to archive the required quality. ______

3.0 CONTRACT REVIEW

12. Do you have procedures to ensure that:

12a. contractual requirements are adequately defined and documented? ______

12b. any requirements differing from those in the tender are resolved? ______

Rat

12c. you have the capability of meeting contractual requirements?

13. Are records of all contract reviews maintained?

14. Are the person(s) responsible for contract reviews identified?

4.0 DESIGN CONTROL (Questions 15-25 for ISO 9001 ONLY)

15. Have you identified the responsibility for each design and development activity?

16. Are these design activities updated as the design evolves?

17. Are the design and verification activities assigned to qualified staff equipped with adequate resources?

18. Are the organizational and technical interfaces between different groups identified?

19. Is the information gathered by the various design groups documented, transmitted (i.e. exchanged) and regularly reviewed?

20. Are design input requirements relating to the product, identified, documented and reviewed for adequacy?

21. Are incomplete, ambiguous or conflicting requirements resolved with those responsible for drawing up these requirements?

22. Are design outputs documented and expressed in terms of requirements, calculations, and analyses?

23. Are characteristics of the design crucial to the safe and proper functioning of the product identified?

24. Do you have person(s) assigned for the verification of designs?

25. Are design reviews recorded and filed?

5.0 DOCUMENT CONTROL

26. Do you have procedures for insuring the identification, documentation, and appropriate review and approval of all design changes and modification?

Rating

27. Are changes to documents reviewed and approved by the same functions/organizations that perform the original review? ______

28. Do you have a master list or equivalent document control procedure to ensure that no outdated or non-applicable documents are used? ______

29. Do you re-issue documents after a practical number of revisions/changes have been made? ______

30. Are all out-dated documents removed from circulation? ______

31. Are the appropriate documents available at all locations where operations essential to the effective functioning of the quality systems are performed? ______

6.0 PURCHASING

32. Are you currently selecting sub-contractors on their ability to meet your quality requirements? ______

33. Do you currently evaluate the quality system of your suppliers? ______

34. Have you established and do you currently maintain records of acceptable sub-contractors? ______

35. Do you review and approve purchasing documents for adequacy of specified requirements prior to release? ______

36. Do you have established procedures for the verification, storage, and maintenance of products supplied by the purchaser? ______

7.0 Purchaser Supplied Product (applies to some suppliers)

8.0 PRODUCT IDENTIFICATION AND TRACEABILITY

37. Is the product identified throughout all stages of production, delivery and installation? ______

38. Do you keep records of product identification? ______

9.0 PROCESS CONTROL

39. Are documented work instructions defining the manner of production and installation written down? ______

40. Are you monitoring production and installation procedures? ______

41. Are procedures available to verify that the assembly process can deliver the product as defined? ______

Ratin

42. In the case of special processes — i.e. processes whose results cannot be fully verified by subsequent inspection and testing of the product- do you monitor and document the process to ensure that specified requirements are met?

10. INSPECTION AND TESTING

43. Do you ensure that incoming product is not used or processed until it has been inspected or otherwise verified as conforming to specified requirements?

44. Are your verification/inspection procedures performed in accordance with the quality plan or other documented procedures?

45. Do you perform in-process inspection and testing according to documented procedures as required by the quality plan?

46. Are non-conforming products identified?

47. Do you perform all phases of final inspections according to documented procedures before final release of the product?

48. When purchased products need to be released for urgent production purposes, are they identified and recorded?

49. Do you maintain records which give evidence that the product has passed inspection and/or test with defined acceptance criteria?

11. INSPECTION, MEASURING, AND TEST EQUIPMENT

50. Are measuring and test equipment periodically calibrated?

51. Are records of calibration maintained?

52. Has the precision and accuracy of all measuring and test equipment been determined?

53. Are calibration records (frequency of calibration, instrument bias, etc.) or other pertinent information labelled on each instrument?

54. Are test hardware or test software periodically checked to prove that they are capable of verifying the acceptability of products released for use?

12. INSPECTION AND TEST STATUS

55. Do you during test procedures use markings, stamps, tags, labels or other such means to identify conforming or nonconforming products?

Rating

56. Do you keep records which identify the inspection authority responsible for the release of conforming product? ______

13. CONTROL OF NONCONFORMING PRODUCT

57. Have you identified areas to isolate nonconforming products? ______

58. Do you keep records of the disposition taken on nonconforming products? ______

59. Are concessions for nonconforming recorded? ______

60. Are repaired and reworked products re-inspected in accordance with documented procedures? ______

14. CORRECTIVE ACTIONS

61. Do you investigate the cause(s) of nonconforming product and the corrective action(s) needed to prevent recurrence? ______

62. Do you analyze processes, work operations, concessions, quality records, service reports, and customer complaints to detect and eliminate potential causes of nonconforming product? ______

63. Do you verify that corrective actions are effective? ______

64. Do you implement and record changes in procedures resulting from corrective action? ______

15. HANDLING, STORAGE, PACKAGING AND DELIVERY

65. Do you maintain procedures for:

65a. Handling ______

65b. Storage ______

65c. Packaging ______

66d. Delivery ______

67. In order to detect deterioration, do you assess at regular intervals the condition of product in stock? ______

16. QUALITY RECORDS

68. Have you established and maintained procedures for the identification, collection, indexing, filing, storage, maintenance and disposition of quality records? ______

69. Do you archive your quality records?

Rating

17. INTERNAL QUALITY AUDITS

70. Do you conduct internal quality audits to verify whether quality activities comply with planned arrangements and to determine the effectiveness of the quality system? ______

71. Are the results of internal audits documented and brought to the attention of the personnel having responsibility in the audited area? ______

72. Are corrective actions on deficiencies found during the audit resolved? ______

73. Are internal audits conducted according to specified procedures? ______

18. TRAINING

74. Do you currently maintain procedures for identifying the training needs of all personnel performing activities affecting quality? ______

75. Do particular jobs require special education/training? ______

76. Do you maintain records of all training activities? ______

19. SERVICING (ISO 9001 ONLY)

77. Have procedures been established for performing servicing? ______

20. STATISTICAL TRAINING

78. Do you currently use statistical techniques to verify the acceptability of process capability and product characteristics? ______

79. Do you rely on any appropriate statistical technique to monitor:

79a. nonconforming products. ______

79b. customer complaints. ______

79c. test and production control. ______

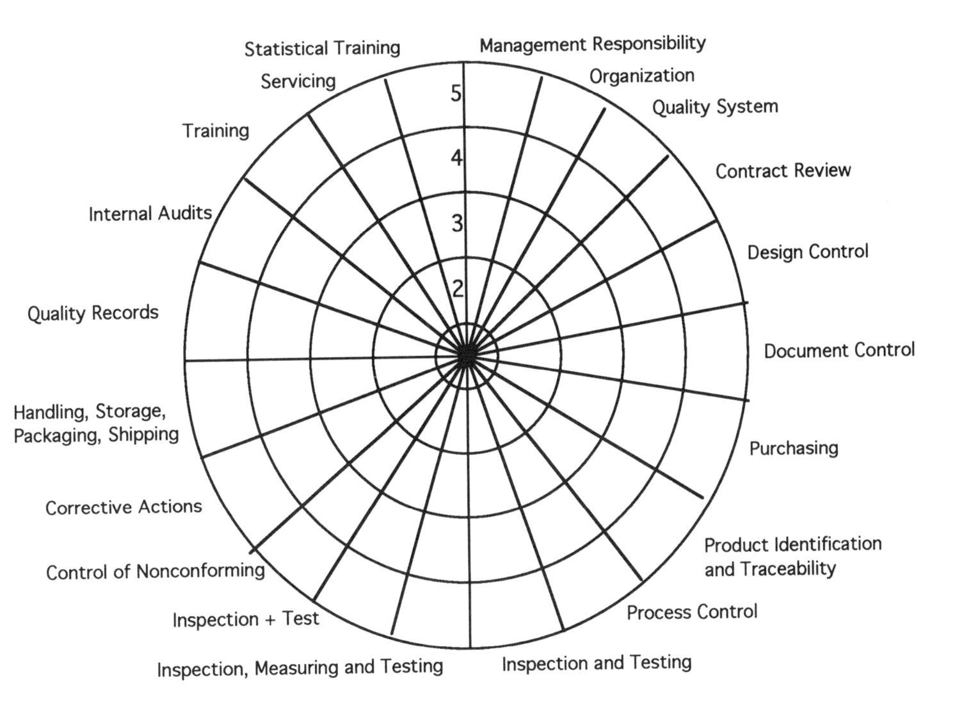

ISO 9000 QUALITY SYSTEM

Index